Geospatial Semantic Web

Chuanrong Zhang • Tian Zhao • Weidong Li

Geospatial Semantic Web

 Springer

Chuanrong Zhang
Department of Geography
Center for Environmental Sciences
and Engineering
University of Connecticut
Storrs
Connecticut
USA

Weidong Li
Department of Geography
Center for Environmental Sciences
and Engineering
University of Connecticut
Storrs
Connecticut
USA

Tian Zhao
Department of Computer Science
University of Wisconsin-Milwaukee
Milwaukee
Wisconsin
USA

ISBN 978-3-319-17800-4   ISBN 978-3-319-17801-1 (eBook)
DOI 10.1007/978-3-319-17801-1

Library of Congress Control Number: 2015942460

Springer Cham Heidelberg New York Dordrecht London

Printed on acid-free paper

Springer is part of Springer Science+Business Media (www.springer.com)

Contents

Chapter 1
Geospatial Data Interoperability, Geography Markup Language (GML), Scalable Vector Graphics (SVG), and Geospatial Web Services

1.1 Geospatial Data Interoperability

As GIS (Geographic Information System) has been widely used by a variety of applications, many geographical databases have been developed by different programs and software. However, it is still a big problem to share these geospatial data and use them for the development of GIS applications. Not that data are not available, there is a huge amount of geographical data stored in different places and in different formats, but data reuse for new applications and data sharing are daunting tasks because of the heterogeneity of existing systems in terms of data modeling concepts, data encoding techniques and storage structures, etc. (Devogele et al. 1998).

There are several commercial desktop GIS software systems that dominate the GIS industry. For example, ESRI ArcGIS, Intergraph GeoMedia, Autodesk AutoCAD, MapInfo Professional, Smallworld GIS, and SuperMap are popular GIS software with high market share. In addition, numerous open-source desktop GIS software systems are also available for geospatial data handling. For example, GRASS GIS, originally developed by the U.S. Army Corps of Engineers, is a popular open-source desktop GIS software system. Other open-source desktop software examples include gvGIS (http://www.gvsig.org/web), JUMP GIS (http://jump-pilot.sourceforge.net/), uDig (http://udig.refractions.net/), SAGA GIS (http://www.saga-gis.org/en/index.html), ILWIS (http://52north.org/downloads/ilwis), MapWindow GIS (http://www.mapwindow.org/), and QGIS (http://qgis.org/en/site/). It is unlikely that all GIS applications will use the same software. Different vendors have their own proprietary software designs, data models, and database storage structures. Thus, geographical databases based on these designs cannot communicate without data conversion. In order to exchange information and share computational geo-database resources among heterogeneous systems, conversion tools have to be developed to transfer data from one format into another. Furthermore, these diverse desktop GIS database structures make remote data exchange and sharing more difficult because of limited accessibility and required data conversion.

C. Zhang et al., *Geospatial Semantic Web,* DOI 10.1007/978-3-319-17801-1_1

Internet GIS or Web GIS creates a unique environment for sharing geospatial data. There are many Internet GIS or Web GIS programs available for data sharing over the Web. For example, Esri's ArcGIS Server, Intergraph's GeoMedia WebMap, MapInfo's MapExtreme, AutoDesk's MapGuide, GE SmallWorld's Internet Application Server, and ER Mapper's Image Web Server are popular commercial Internet GIS programs. Although these Internet GIS programs offer better tools for data sharing over the Web, they also have the problems of proprietary software designs, data models, and database storage structures. Sharing of data, facilitated by the advances in network technologies, is hampered by the incompatibility of the variety of data models and formats used at different sites (Choicki 1999).

There are two problems resulting directly from the non-interoperability of databases. One is the change in data accuracy. After data are converted from one format to another, problems such as coordinate precision, errors of omission, missing or wrong attribute names, and incorrect topology may occur (Noronha 2000). For example, the spatial features with topological errors in *Coverage* format might be removed after converting from *Coverage* format to *Shapefile* format. Data conversion may create more map errors and uncertainty for many practical applications. The second problem is the investment of time and money for data conversion. A lot of money and time has been wasted on data conversion or developing data conversion tools. Most investments by today's GIS users lie in three areas: data conversion, development of application specific extensions to general purpose GIS products, and learning of applications of software and data to enhance productivity. Among these three areas, data exchange and conversion account for a very high percentage (Siki 1999).

Data interoperability means the ability to utilize a range of data formats. Interoperable geospatial data can be used by different types of programs and applications. With interoperable geospatial data, users should be able to request, access, and integrate data easily no matter where the data are stored (locally or remotely). Geospatial data interoperability is extremely important for geospatial applications, because large amounts of spatial data of different geographical formats existed and there are increased demands for re-use of these existing spatial data for decision-making. Interoperability of geospatial data eliminates barriers for data sharing and allows users to directly access, map, visualize and analyze data with different spatial data formats. Interoperable geospatial data makes fast information delivery and sharing across departments possible.

How to realize the goal of data interoperability? There are two approaches to data interoperability—database integration and standardization (Devogele et al. 1998). Database integration is a very sophisticated approach. A very basic approach is to provide users with a global catalogue of accessible information sources, where each source is described by associated metadata, including representation mode, scale, last update date, and data quality level, among others (Stephan et al. 1993). Current database integration has evident drawbacks related to lack of scalability, consistency, and duplication (Devogele et al. 1998). The second approach to interoperability is through standardization. The definition of standard data modeling and manipulation features provide a reference point which facilitates data exchange among heterogeneous systems (Devogele et al. 1998).

In literatures, several useful standards have been developed to facilitate spatial data exchange. For example, Geographic Data File (GDF) is an interchangeable file format for geographic data sharing. GDF is specifically designed for spatial data exchange for Intelligent Transportation Systems (ITS). It defines a set of spatial features, attributes, and relationships that are particularly relevant to ITS applications, and specifies a set of useful data structures and data formats. This makes it readily usable for off-line data exchange. For another example, Spatial Data Transfer Standard (SDTS) is a general purpose standard that is flexible and adaptive (NIST 1994). The purpose of SDTS is to promote and facilitate the transfer of digital spatial data among different computer systems, while preserving information meaning and minimizing the need for information external to the transfer. With anticipated extensions and refinements, SDTS was expected to become an important data format for ITS spatial data transfer or a neutral format for data archiving (Arctur et al. 1998).

However, currently both GDF and SDTS were not so widely used as originally anticipated. Despite the existence of an ISO GDF standard, in practice, GDF files are not fully interchangeable due to vendor specific extensions. There are interoperability problems among GDF map products from different vendors. Several barriers block the popularity of SDTS. These barriers include the complexity of SDTS, slowness in the development of practical SDTS profiles, restriction of each SDTS dataset to a single profile, lack of a clear definition of geospatial features in SDTS, and ambiguity in the means of specifying cardinality of relationships in a data model (Arctur et al. 1998).

The creation of a new standard data exchange format, Geography Markup Language (GML), represents another important step taken by the geospatial community towards geospatial data interoperability. In next section, we will introduce this new standard data exchange format GML.

1.2 Geography Markup Language

Geography Markup Language (GML) is "an XML grammar written in XML Schema for the modeling, transport, and storage of geographic information including both spatial and non-spatial properties of geographic features" (OGC 10-129r1 2012). It is developed as an implementation specification by the Open Geospatial Consortium (OGC) to foster data interoperability and exchange between different systems over the Internet.

The GML model contains a set of primitives that can be used to develop application specific schemas or application languages. These primitives include

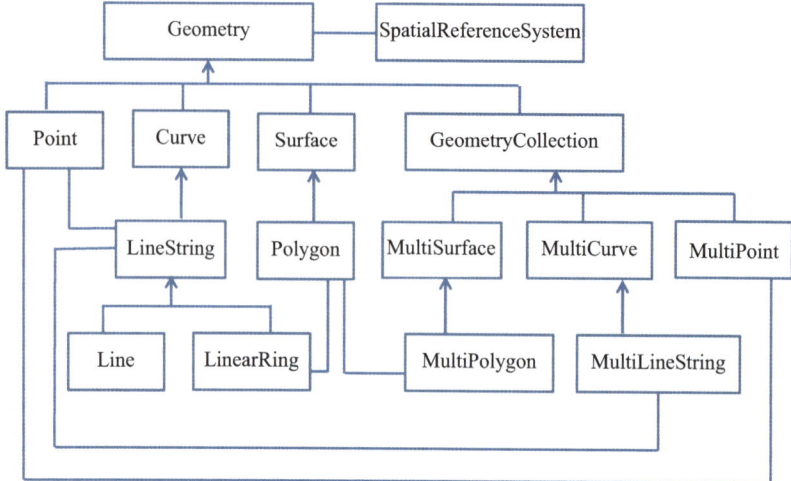

Fig. 1.1 The geometry model for simple features in GML 3.2

Feature, Geometry, Coordinate reference system, Topology, Time, Dynamic feature, Coverage (including geographic images), Unit of measure, Directions, Observations, and Map presentation styling rules. GML is based on ISO 19100 series of International Standards and the OpenGIS Abstract Specification (http://www.opengeospatial.org/standards/as) that model the world in terms of *features.* *Features* are different from *geometry objects.* A feature is an abstraction of a real world phenomenon and it can represent a physical entity such as a building or a road. A *feature* may or may not have geometric aspects. A *geometry object* can be any real-world object that is associated with a location or region instead of a physical entity, and it is different from a feature. The key GML *geometry object* types in GML 1.0 and 2.0 include *Point, LineString,* and *Polygon.* The simple features in GML 3.2 include more types. Figure 1.1 illustrates the geometry model for simple features in GML 3.2. From the figure it can be seen that the geometry types not only include *Point, Curve (LineString), Surface (Polygon),* but also include *MultiPoint, MultiCurve, MultiSurface, and MultiGeometry.* In addition, GML 3.0 and higher versions also include a *"coverage"* structure, which is used for the *"raster"* data such as remotely sensed satellite imagery. Thus, GML 3.0 and higher versions are able to represent real-world phenomena using more complex feature types.

Each feature attribute and feature association role in GML can be encoded as a property of a *feature* in an XML element. The following codes provide an example to encode a *feature* (*School* Feature) with *Point* geometry and some other attributes of the feature:

```
<School>

    <SchoolName>Mansfield Middle School</SchoolName>

    <history>42</history>

    <BuildingName> Audrey P. Beck Municipal Building </ BuildingName >

    <TownName>Town of Mansfield</TownName>

    <gml:location>

            <gml:Point>

                    <gml:coord><gml:X>1.0</gml:X><gml:Y>1.0</gml:Y></gml:coord>

            </gml:Point>

    <gml:location>

</School>
```

GML offers standard ways to describe these spatial features and their corresponding properties in terms of GML Schema, including schemas to describe features, coordinate reference systems, geometry, topology, time, units of measure, and generalized values. In fact, these schemas conform to the ISO standards, thus they allow GML applications that follow these schemas to be interoperable. The ISO standards include:

- ISO/TS 19103—Conceptual schema language,
- ISO DIS 19107 Geographic Information—Spatial Schema,
- ISO DIS 19108 Geographic Information—Temporal Schema,
- ISO 19109—Rules for application schemas,
- ISO DIS 19111—Spatial referencing by coordinates,
- ISO DIS 19123—Schema for coverage geometry and function,
- and ISO 10148—Linear Referencing.

The correspondence between GML object types and classes in these ISO 19100 series of International Standards and the OGC Abstract Specification can be mapped in a table. In addition, GML also provides XML encodings for additional concepts that are not modelled in the ISO 19100 series of International Standards or the OpenGIS Abstract Specification. For example, moving objects are encoded in GML but they are not defined in the ISO 19100 series of International Standards. Further, applications can design their own GML application schemas, which may extend or restrict the types defined in the standard GML schema. Application schemas may use directly non-abstract elements, attributes, and types in the standard GML schema.

In general, GML schemas reconcile the need for standardization with the need for diversity by providing a standard means of extending the GML format. The direct consequence of applying schemas with GML is that organizations can define formats to suit their needs and exchange information without developing translators for those formats. This has impacts on both the cost and risk of exchanging data.

The following codes provide an example of the application schema fragment developed for the above example of encoding *School* Feature with point geometry:

```
<element name="School" type="ex:SchoolType" substitutionGroup="gml:_Feature"/>

<complexType name="SchoolType">

    <complexContent>

        <extension base="gml:AbstractFeatureType">

            <sequence>

                    <element name="SchoolName" type="string"/>

                    <element name="history" type="integer"/>

                    <element name="BuildingName" type="string" minOccurs="0"

                        maxOccurs="unbounded"/>

                    <element name="TownName" type="string" minOccurs="0"

                        maxOccurs="unbounded"/>

                    <element ref="gml:location"/>

            </sequence>

        </extension>

    </complexContent>

</complexType>
```

Overall, GML provides an open and vendor-neutral framework for the description of geospatial application schemas for the transport and storage of geographic information in XML (W3C XML-1 2004; W3C XML-2 2004). A large number of XML elements and attributes in GML support a wide variety of capabilities and can be used to encode dynamic features, spatial and temporal topology, complex geometric property types and coverages (OGC 07-036 2007). Because XML is a universal format for structured documents and data on the Web, it is easy to transform. Using XSLT or almost any other programming language (VB, VBScript, Java, C++,

Javascript), users can transform XML from one form to another. By adhering to an open, non-proprietary standard, GML documents can be manipulated, transformed and presented in the same flexible way as XML contents (OGC 10-100r3 2011).

Because GML data are stored in plain text, which is vendor-neutral, information stored in GML is not locked into a proprietary binary format. GML can be readily integrated with a wide variety of non-spatial data types including text, business transactions, graphics, audio, voice and more. This capability would greatly enhance the value and accessibility of geospatial information. For example, users can easily insert a map in a financial report, or vice versa (Peng and Tsou 2003). In addition, as a text format, GML can be easily transmitted across a variety of platforms over the Internet. Thus GML enables disparate systems to share information easily.

The linking mechanism of HTML (one web page linking to another), is one of the key foundations of the Web. GML goes further than HTML by providing *XLink* and *XPointer* mechanisms for linking multiple distributed resources into a complex association. *XLink* is used in GML to implement associations between objects by reference and is a standard method to support hypertext referencing in XML. *Xpointer* can point to a resource within or outside the same XML document. Through *XLink* and *XPointer*, different features and feature collections, which may be located remotely, can be associated together at the feature level over the Web.

Finally, GML has been designed to uphold the principle of separating content from presentation. Therefore, GML only offers mechanisms for encoding of geographic features and it does not develop ways to visualize data. However, because GML is an XML application, it can be easily styled into a variety of presentation formats for visualization. For example, GML can be directly styled into SVG (Scalable Vector Graphics) for generation of graphical maps. We will introduce SVG in the following section.

1.3   Scalable Vector Graphics

Scalable Vector Graphic (SVG) is a XML-based standard developed by the W3C (World Wide Web Consortium) to describe two-dimensional graphics, especially for display in the Web browser. As the name indicates, SVG is a vector graphic, which is different from the raster image formats such as GIF, JPEG, and PNG. As a vector graphic SVG uses mathematical statements to describe the shapes and paths of an image. Please note that a raster graphic uses a grid of x and y coordinates to describe or display pattern information in monochrome or color values on a display space.

SVG has three types of graphic objects: vector graphic shapes, images, and texts. Vector graphic shapes represent some combination of straight line and curve. Images represent an array of values that specify the paint color and opacity at a series of points on a rectangular grid. Texts represent some combination of character glyphs. These graphical objects can be grouped, styled, transformed and composited into

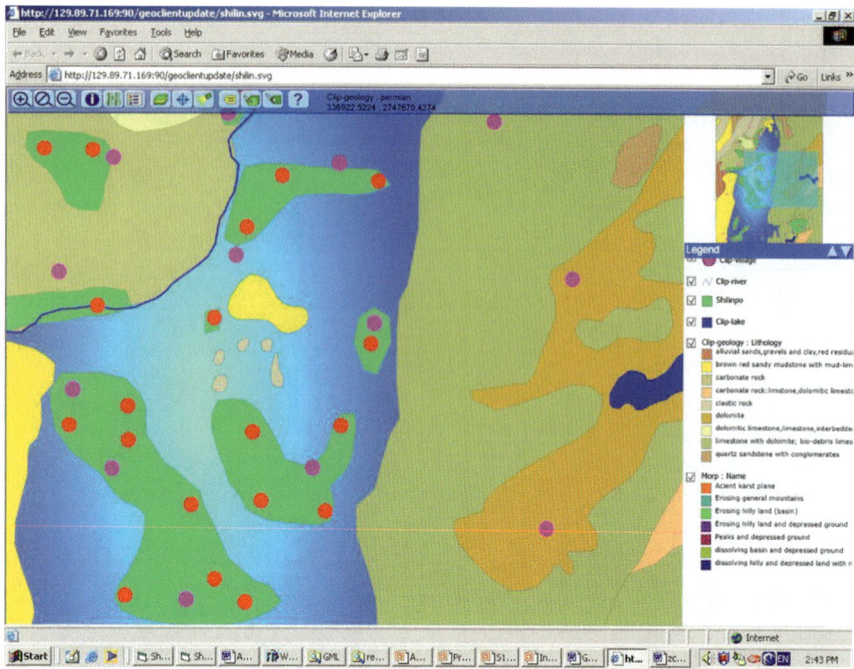

Fig. 1.2 An example of SVG maps

different rendered objects. Vector graphics in SVG include geometric objects such as points, lines, curves, and polygons.

Compared with the raster format which has to store information for every pixel of a graphic, vector graphics have greater flexibility—in fact, SVG can integrate raster images and can combine a raster image with vector information. Unlike most existing XML grammars, which represent either textual information or raw data such as attribute information, SVG not only provides rudimentary graphical capabilities but also has more capabilities than raster images. SVG provides a rich, structured description of vector and mixed vector/raster graphics and can be used stand-alone, or with other XML grammars.

The use of the word "scalable" in SVG has two meanings. First, vector graphic images can easily be made scalable, i.e., not being limited to a single and fixed pixel size. This means SVG format can be displayed on any device of any size (whether a cell phone or a 19-inch computer monitor) and any resolution without changing the image clarity. This contrasts with raster image files, which are difficult to modify without loss of information. Figure 1.2 shows an example of SVG on the screen of a high resolution computer. This example can be printed out using the full resolution of a high-resolution printer. In addition, SVG can let users zoom in on any portion of the GML data without any degradation in the quality of maps (Zhang et al. 2003). As shown in Fig. 1.3, no matter how many times of zooming in, a resolution-independent high quality map can always be provided. Of course,

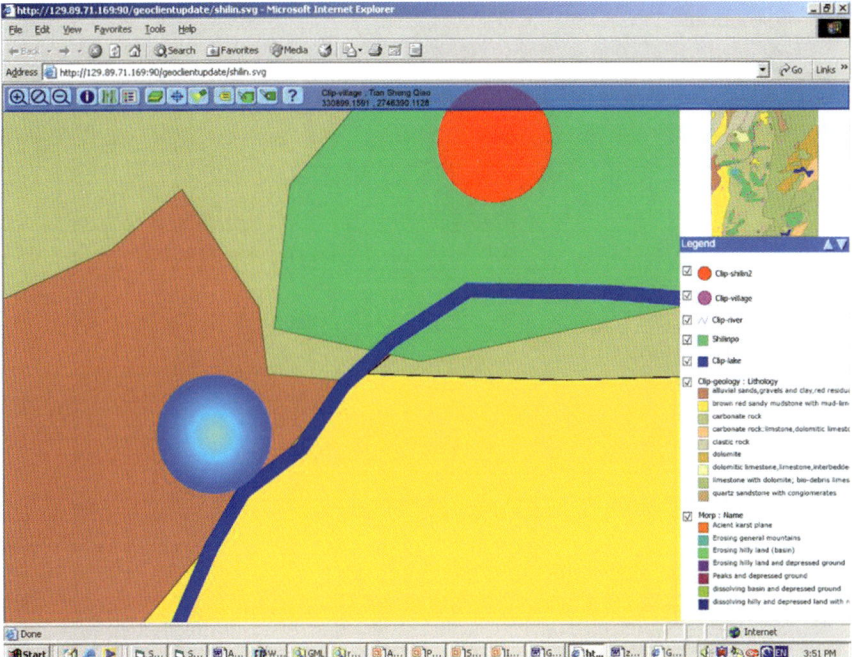

Fig. 1.3 High quality graphics when zooming

the quantity of information in SVG is still restricted by the original data. However, the SVG vector maps never have staircase effects as users can see when printing an enlarged pixel-based GIF and JPEG image. So this has solved the blur problem of raster image maps when a scale is changed (especially amplified).

Second, "scalable" means that the technology can grow to a larger number of files, a large number of users, and a wide variety of applications on the Web (W3C 2001). The same SVG graphic can be placed at different sizes on the same Web page, or re-used at different sizes on different pages (W3C 2001). An SVG file also can be referenced or included inside other SVG graphics. Different applications or programs may build their maps by using parts of SVG.

Other characteristics of SVG include a smaller file size and searchable text information. An SVG file is usually smaller than a raster file for the same map resolution. Thus it can be transferred across the Internet more quickly. Text information inside SVG is still text and can be searchable, while text information inside the raster file becomes integrated into the image and is no longer recognized as text. SVG is particularly suitable for displaying intelligent maps, because geometric objects such as points, lines, and polygons are identifiable. Raster images on the other hand contain information about every pixel. Thus points, lines and polygons in raster images are no longer recognizable. Users can directly work with spatial features on an SVG but cannot on a raster graphic image. Users can query a SVG map. Figure 1.4

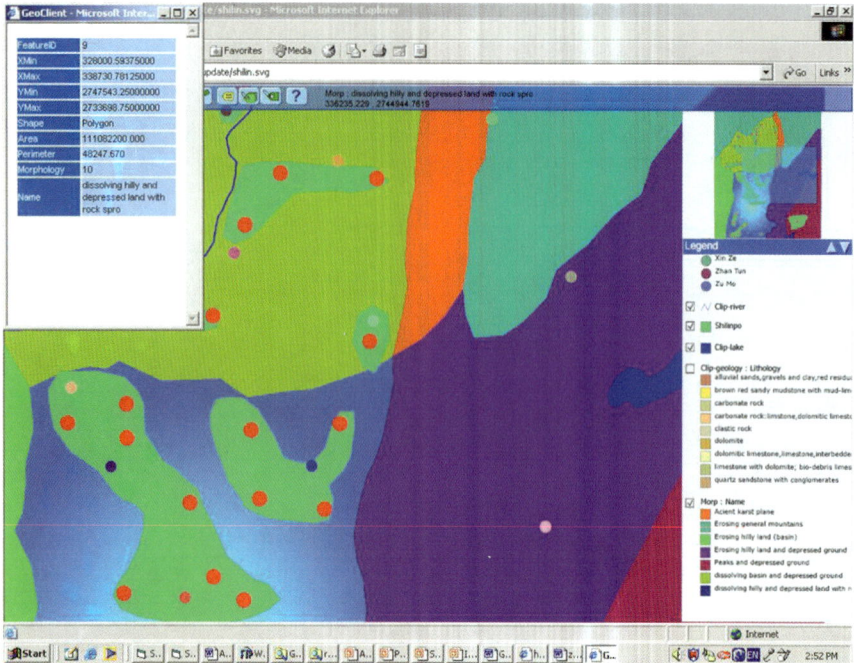

Fig. 1.4 Querying an SVG map

illustrates an example of querying an SVG map—when users click on a feature in a SVG map, the related attribute table for the feature can be brought out.

SVG is based on XML and, therefore, conforms to other XML-based standards and technologies, such as XML *Namespace*, *XLink*, and *XPointer*. *XLink* and *XPointer* allow for linking from within SVG files to other files on the Web (W3C 2001). As a W3C standard, SVG can integrate with other W3C specifications and standards efforts.

For the development of Internet GIS, SVG has the potential to play an important role for three reasons (Peng and Zhang 2004). First, it reduces the size of a map image by allowing complex scalable cartography in a highly compressed form. Second, as an XML application, SVG provides hyperlinks to many other files and vector and raster graphics. It can also work directly with other XML-based technology. Third, since an SVG file is an XML file, it offers superior portability (Peng and Zhang 2004). That is, an SVG file could be edited and displayed in any environment regardless of computer operating systems and Web browsers. The combination of these three characteristics means that SVG can play an important role in the development of Internet GIS.

GML data can be transformed into the SVG format by using an XSLT processor through combination with a Style sheet. XSLT (Extensible Stylesheet Language Transformations) is a language for transforming XML documents into other XML

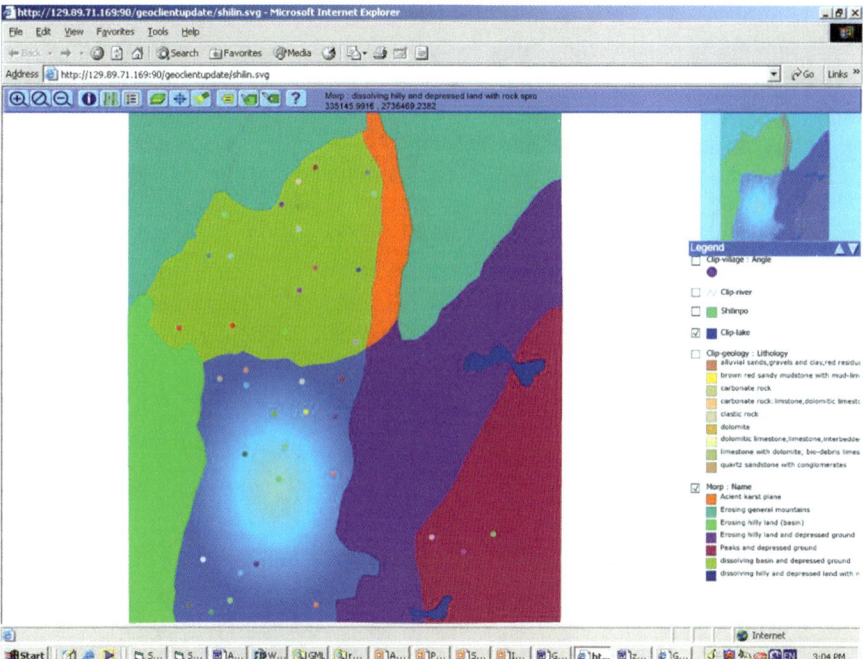

Fig. 1.5 An example of a styled SVG map from a GML data set

documents, or other objects such as HTML for web pages. Currently there are several implemented XSLT Processors. For example, *Altova RaptorXML Server* is a cross-platform engine that supports XSLT 1.0 and 2.0. For another example, the *libxslt* is a free library released under the MIT License. Among these implemented XSLT processors, *Xalan* (Apache 2006) and *Saxon* (Kay 2013) are two popular ones. With the help of these XSLT processors, GML data can be transformed or "styled" into an SVG file for display using a Style sheet. Figure 1.5 illustrates an example of a styled SVG map from a GML data set.

Although the graphics created by SVG can scale nicely to different resolutions and sizes, SVG has limitations. For example, although SVG allows flexibility in uniformly scaling an entire graphic, there is little support for scaling only parts of the graphic (Badros et al. 2001). Differential scaling can allow users to enlarge a part of a map while simultaneously reducing the size of the other parts of the map. The result would be that the parts not enlarged become smaller in size to accommodate the area that is enlarged. With the differential scaling, the areas of interest can be viewed in greater details while preserving the relationships amongst all parts. Thus differential scaling should be very useful for many geospatial applications. However, the current SVG implementation cannot support the differential scaling capability. Another limitation example of the current SVG implementation is semantic zooming. Semantic zooming means that zooming will preserve the semantic presentation but change its appearance (Badros et al. 2001). Semantic preserving manipulations give users the ability to change the layout of the map objects inter-actively, while preserving the logical structure or semantic meaning. However, se-mantic zooming is not supported in current SVG implementations. Currently efforts are being made in the geospatial community to overcome these limitations of SVG and make it better for visualizing complex heterogeneous spatial-temporal data sets.

While GML provides a means to encode and transport geospatial features into XML, SVG provides a means to display these GML-coded geospatial features into vector maps on the Web. One issue of concern is how to conduct queries and extract features from a database to respond to user requests. The Geospatial Web Services Implementation Specifications developed by the OGC serve this role. We will intro-duce the OGC Web Services in the following section.

1.4   Geospatial Web Services

With the development of open standards, web services emerged for data interoper-ability over the Web. Web services are self-contained and self-described software components that can be discovered and invoked by other software components through the web. Web services can be seen as "Evolutionary" of the Web because they make the computer systems shift from the client-server architecture to the peer-to-peer architecture, which makes distributed systems possible. In the web service view, every different system or component offers some services for others, and every system does its job by just calling or combining suitable services over the Internet (Cömert 2004). Through a web service, a collection of functions or tools are packaged as a single entity for use by other applications and programs over the Web.

Because web services are based on standard protocols such as SOAP (Simple Object Access Protocol), any web service can incorporate other web services for interoperability. However, web services can be written in any language. The emer-gence of web services provides the interoperable capability of cross-platform and cross-language functionality in the distributed net environment (Jia et al. 2004; An-derson and Moreno 2003). Because web services communicate with each other us-ing HTTP (the Hypertext Transfer Protocol) and XML, any device including wire-less-network computers or cellar phones can host and access web services. Because the concepts behind web services are easy to understand, developers can quickly create or deploy web services for their applications. Currently, almost all of the major vendors support web services technology.

Within the broader context of web services, OGC web service specifications deal with geographic information on the Internet. OGC web services are evolutionary web standards that enable integrations of different online GIS data and location informa-tion. With OGC's web service specifications and technologies, users can "wrap" ex-isting heterogeneous spatial data into a web service and enable many potential clients to use the service (OGC Interoperability Program White Paper 2001). OGC web services can be treated as a "black box" to perform a task by dynamically connecting interoperable service chains for different applications (OGC White Paper 2001).

The major OGC web service specifications include Web Feature Service (WFS) Specification, Web Map Service (WMS) Specification, Web Coverage Service (WCS) Specification, Web Processing Service (WPS) Specification, and Catalogue Service (CS) Specification. The Web Feature Service is an implementation specification (OGC 04-094 2005), which allows a client to retrieve, query, and manipulate feature-level geospatial data encoded in GML (Geography Markup Language) from multiple sources. The Web Map Service is capable of creating and displaying maps that come simultaneously from multiple heterogeneous sources in a standard image format (OGC 0-042 2006). The Web Coverage Service provides access to potentially detailed and rich sets of geospatial information in forms that are useful for client-side rendering, multi-valued coverage, and input into scientific models and other clients (OGC 09-110r3 2010). The Web Processing Service defines rules for standardizing inputs and outputs (requests and responses) of geospatial processing services (OGC 05-007r7 2007). The Catalogue Service provides catalogues for OGC web services and supports the ability to publish and search collections of descriptive information (metadata) for data, services, and related information objects (OGC 07-006r1 2007).

1.4.1 Web Feature Service

The Web Feature Service specification is an OGC specification for describing data manipulation operations at the feature level on OGC's simple features (e.g., points, lines, and polygons) (OGC 09-025r1 2010). The Web Feature Service is written in XML (Extensible Markup Language) and uses an open-source standard GML to represent features.

Web Feature Service (WFS) allows a client to retrieve geospatial data encoded in GML from multiple data sources. Because GML uses the geospatial feature model and represents geospatial features, WFS allows access and exchange feature-level geospatial data. Feature-level geospatial data sharing overcomes problems caused by web data sharing at the file levels. In fact, most prior research and professional practices have been focused on web data sharing at the file level. That is, to share and exchange geospatial information users must request entire datasets or data files from different data sources via online downloading (Zhang and Li 2005).

There are several problems with the file-level data sharing systems (Zhang and Li 2005). First, file-level data sharing usually requires data integration techniques such as data conflation due to differences in semantics, data model, and data format. Data conflation is a tedious, subjective, and often error-prone process for consolidating differences between two or more data files. Second, data updated from one source at the file level cannot be automatically propagated to other related data or applications. Data sharing at the file level usually causes latency of data updates,

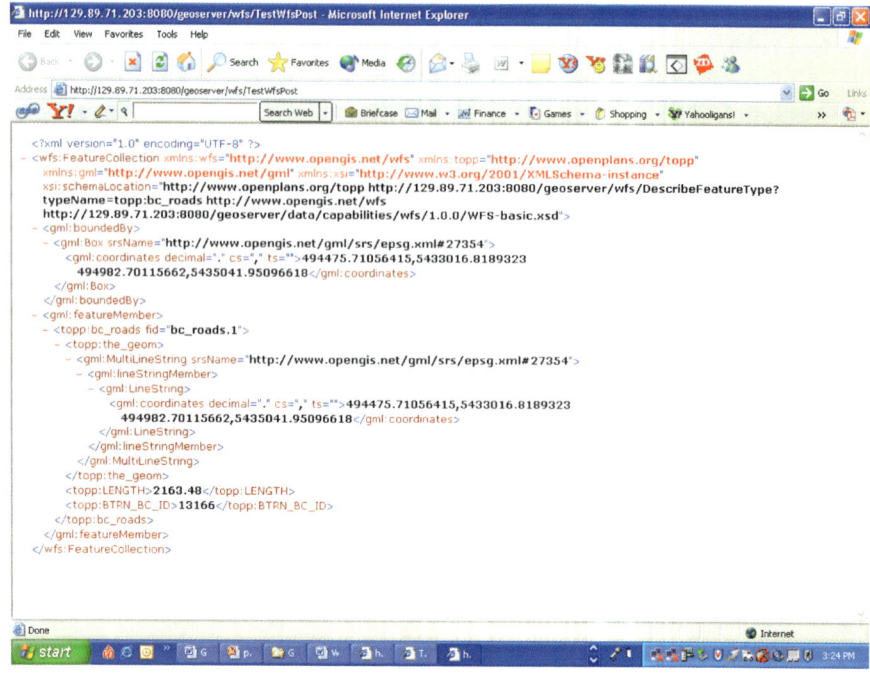

Fig. 1.6 Querying one route segment in GML format over the Web via WFS

which causes problems for time-critical applications that need real-time data access to the most up-to-date information. Third, file-level data sharing makes it difficult to provide feature-level data search, access, and exchange in real time over the Web. Applications must download a whole data file for analysis, even if they need only several features of the file or have interests only in a small area of the file. Downloading entire datasets or data files will increase the time of data acquisition and analysis and affect the speed of decision-making. Therefore, although file-level data sharing and data integration are useful, they are insufficient to meet the demand of the applications that need real-time access and exchange of the most up-to-date feature-level data. WFS overcomes these problems by allowing users to share geospatial data over the Web at the feature level. Figure 1.6 shows an example of querying one route segment feature in GML format over the Web using WFS. Figure 1.7 shows the corresponding SVG map displayed in the Web browser client. Figure 1.8 illustrates the results of querying and integrating several features in a particular geographic area from two WFS data servers.

The proprietary systems' support of GML in WFS is through a *DataStore*. The *DataStore* is used to communicate with WFS servers. The *DataStore* can read and write to a WFS server. It can parse and write WFS requests. A *DataStore* in fact represents a physical source of geospatial data. The *DataStore*, which is the WFS interface for accessing spatial data of different formats, can transform a proprietary data format such as ESRI's Shapefiles into the GML feature representation. The

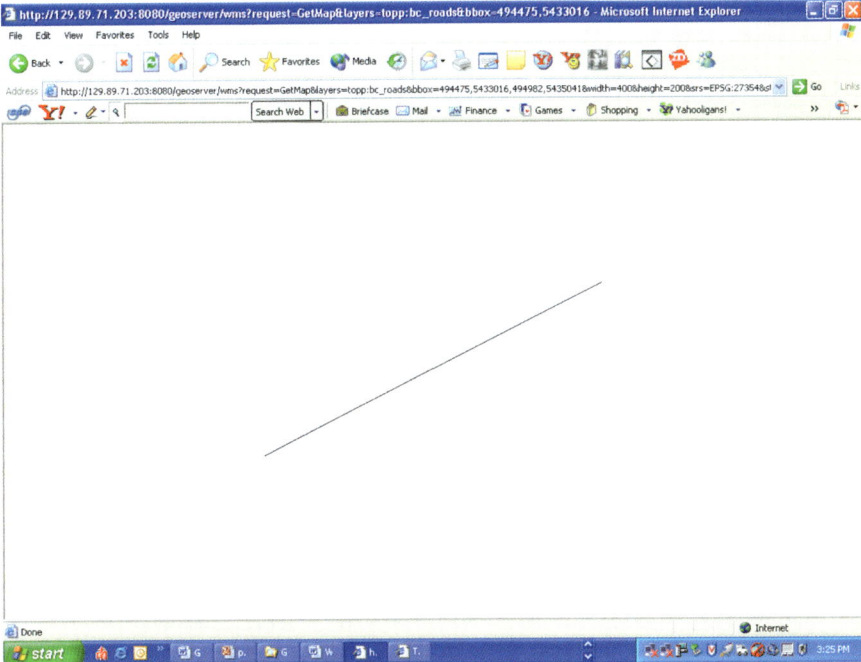

Fig. 1.7   The corresponding SVG graphic map of the queried route segment feature

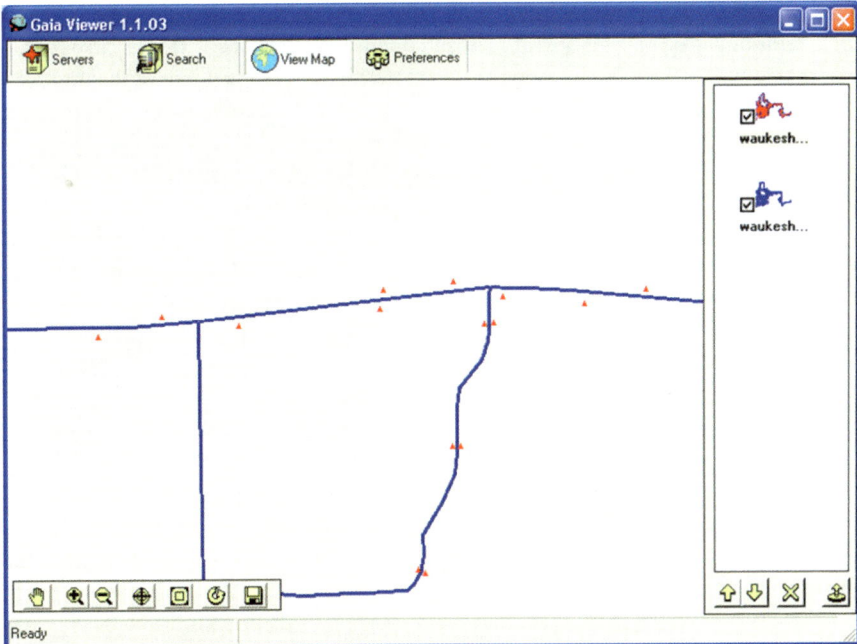

Fig. 1.8   Querying and integrating several features from two WFS data servers

Fig. 1.9 Simplified WFS architecture. (Sources: OGC document 04-094 2005)

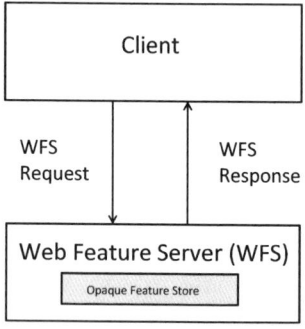

DataStore can be seen as a wrapper for the GML parsers and data writers. The original databases in a DataStore can be in any format, but the different formats of databases are opaque to client applications. A *DataStore* consists of one or more *FeatureTypes* or data layers, which may be a table in a database, a geospatial data file (such as a Shapefile), or a directory in a VPF (Vector Product Format) library. In addition, a DataStore is able to read or write from a URL.

The data retrieval process via a *DataStore* using WFS is shown in Fig. 1.9. Users send requests in XML to a WFS server; the WFS server, which connects with databases with various formats, processes the requests and sends the responses in XML back to users. With the *DataStore*, users do not need to define connection parameters for each table in a large database.

The WFS provides interfaces for five basic data manipulation operations on GML features (OGC 09-025r1 2010): *discovery* operations, *query* operations, *locking* operations, *transaction* operations, and operations to *manage stored parameterized query expressions*. These operations are performed independent of the underlying *DataStore*. Using these operations, users can access, query, create, update, and delete GML features over the Web from WFS servers.

Discovery operations determine the capabilities of WFS through the "*GetCapabilities*" action and retrieve the application schema through the "*DescribeFeatureType*" action. The "*GetCapabilities*" action provides a document describing a WFS service provided by a server. The "*DescribeFeatureType*" action describes the feature types offered by a WFS instance such as how a WFS anticipates feature instances to be encoded on input and how feature instances to be encoded on output.

Query operations retrieve features or values of feature properties from the underlying *DataStore* through the "*GetFeature*" action and the "*GetPropertyValue*" action. The *GetFeature* action returns users a selection of features from a *DataStore* by using the query expressions specified in the user request. The "*GetPropertyValue*" action brings a set of properties of the features in a *DataStore*. The value of a property may be a literal (such as a number or text), or may be structured using XML elements (e.g., the value of a property of a feature may be that of another feature).

Locking operations permit exclusive access to features for modifying or deleting features through the "*GetFeatureWithLock*" action and the "*LockFeature*" action.

They guarantee that while a feature is being modified by one user, another user cannot come along and update the same feature in the database. In other words, while one transaction accesses a data item, no other transaction may modify the same data item. The *LockFeature* action enables a *long-term feature locking* mechanism to ensure consistency. The *GetFeatureWithLock* action is functionally similar to the *GetFeature* action. The difference is that a WFS with the *GetFeatureWithLock* action will not only give a response document similar to that of the *GetFeature* action but will also lock the features in the result set.

Transaction operations allow users to create, modify, replace, and delete features in a WFS *DataStore*. The capability to create, delete, and update features over the Web provides the potential to conduct spatial analysis, modeling, and other operations based on spontaneous access to distributed geospatial data at the feature level. For example, a transportation department staff can instantly add a new bus stop feature to their remote database by using the creating capability of WFS over the web.

Stored query operations allow users to create, drop, list, describe, and change the parameterized query expressions through the *"CreateStoredQuery"*, *"ListStoredQueries"*, *"DescribeStoredQueries"*, and *"DropStoredQuery"* actions. The *CreateStoredQuery* action creates the stored queries for the WFS system. The *ListStoredQueries* action lists the stored queries available at a WFS server. The *DescribeStoredQueries* action gives detailed metadata about each stored query expression that a WFS server provides. The *DropStoredQuery* action allows previously created stored queries to be deleted from the WFS system.

Depending on operations implemented, WFS servers can be grouped into four categories: Simple WFS server, Basic WFS server, Transactional WFS server, and Locking WFS server. A Simple WFS server only implements several basic operations such as *GetCapabilities*, *DescribeFeatureType*, *ListStoredQueries*, *DescribeStoredQueries*, and *GetFeature* with only the *StoredQuery* action. A Basic WFS server implements the same operations as those for a Simple WFS server, in addition to implementing the *GetFeature* with the *Query* action and the *GetPropertyValue* operation. A Transactional WFS server implements all the operations in a Basic WFS server plus the *Transaction* operation. A Locking WFS server implements all the operations in a Transactional WFS server plus one of the *GetFeatureWithLock* or *LockFeature* operations.

In general, the feature-level data manipulation of WFS allows users to download only those feature data that they are interested in instead of the entire dataset; this makes WFS valuable for many applications, especially the time-critical applications. Because WFS allows users to access the specific data needed at the feature level from distributed sources, it can largely reduce the time spent on geospatial data acquisition and integration. Figure 1.10 illustrates that a fireman may download only several geographical features close to a fire incident with WFS, rather than download the entire dataset without WFS.

The *XLink* in GML from WFS can link or associate spatial features from different sources, so that the update in one data source from WFS can be immediately reflected in, or propagated to, other related data sources or applications from WFS. Figure 1.11 illustrates an example of *XLink* functions from WFS. In this example,

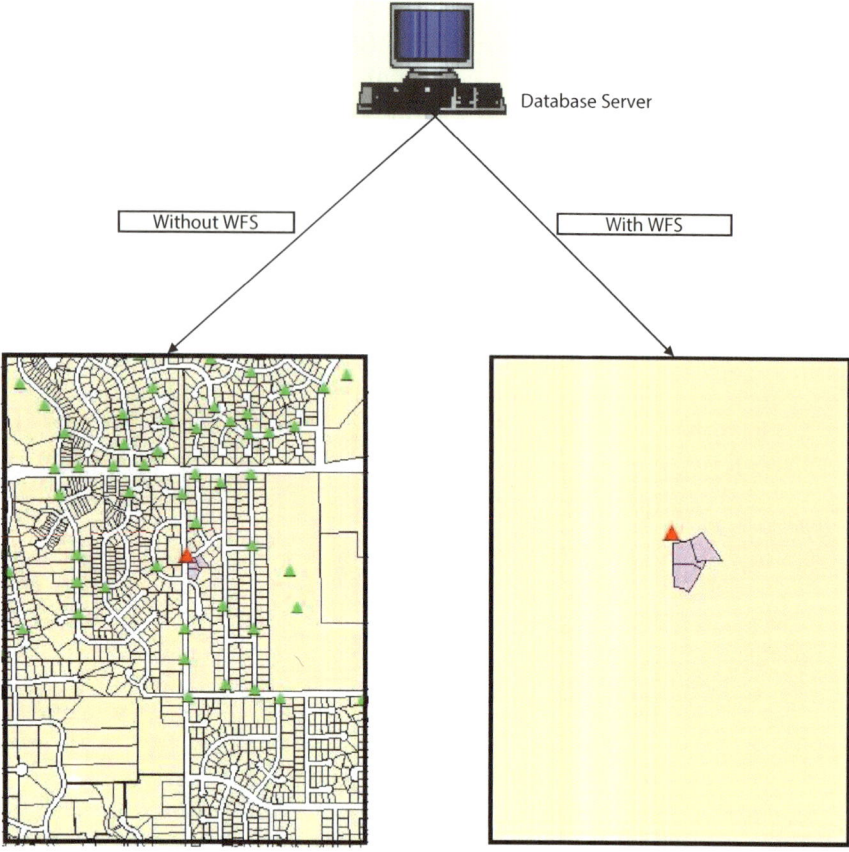

Fig. 1.10 Downloading multiple features with WFS. (Sources: Zhang and Li 2005)

city data are associated with *street*, *building*, *lake*, and *facility* data in different WFS servers via *XLink*. Through the *XLink* function, when Department 1 in the city updates its *street* data in its WFS server, the related *city* data from the WFS sever in the City Government Office will also be automatically updated. For example, When Department 1 adds a new street in its database (red color in Fig. 1.11) via WFS, the database from the WFS sever maintained by the City government office will also be automatically updated because its data is associated to the street data in the WFS sever at Department 1 by *XLink* over the web. This can avoid latency of data update caused by data sharing at the file level. Within a file-level data sharing system, data updated in one department usually cannot be made available immediately to other departments, because data file delivery or downloading is infrequent and users from other departments have to identify and update the changed data manually after obtaining the data files. Through the *XLink* technology, WFS makes automatic and immediate data update at the feature level possible over the Web.

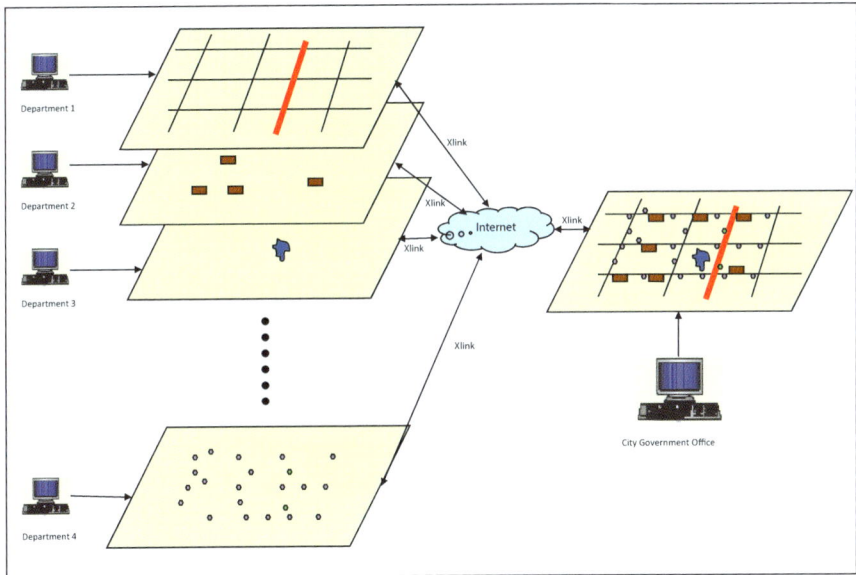

Fig. 1.11 An example of *XLink*. (Sources: Zhang and Li 2005)

1.4.2   Web Map Service

The OGC Web Map Service (WMS) is based on its WMS specification and the ISO/ TC211 specification (ISO 19128). Web Map Service is capable of creating and displaying maps that come simultaneously from multiple sources, which may be both remote and heterogeneous, in standard image formats such as PNG, GIF or JPEG, or sometimes as vector-based SVG or Web Computer Graphics Metafile (WebCGM) formats (OGC 06-042 2006). Web Map Services can be invoked through submitting URLs (Uniform Resource Locators) requests, which provide map and coordinate reference system information. Common for all URLs is that they require a version number and a parameter telling the server which of the request types is chosen. The URLs may also contain other information such as contents, location, and coordinate reference system of the requested map, and the size and format of the output image. For example, the following URLs can be used to extract a map from a WMS server:

http://gis.geog.uconn.edu:8080/geoserver/wms?request=GetMap&layers=topp:ct_streets&b

box=489153,5433000,529000,5460816&width=400&height=200&srs=EPSG:27354&styles

=normal&Format=image/png.

The URLs of this example indicate the following information:

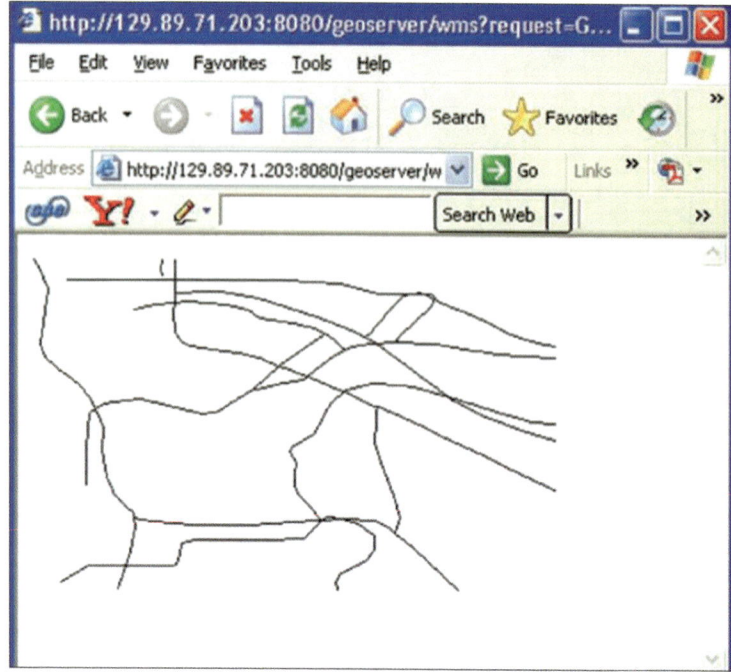

Fig. 1.12 A requested map through the WMS over the Web. (Source: Zhang and Li 2005)

- The requested map comes from the "streets" layer of the "ct" DataStore in the "topp" namespace;
- The EPSG code of its coordinate reference system is 27354;
- Its style is normal (a detailed information about style will be introduced later); and
- The format of output image is PNG.

Note that a layer in WMS generally refers to a single file or a table in a database and it may represent a number of features that have the same type of geometric and non-geometric properties. A *DataStore* in WMS is similar to a *DataStore* in WFS. A *DataStore* in WMS is a single physical source of geographic data, and it may have one or more than one feature layers. Like other OGC web services, *Namespace* in WMS is used to discriminate XML vocabularies from one another and it needs to be a unique identifier (use Uniform Resource Identifiers (URI)). Figure 1.12 illustrates a requested map through the WMS over the Web through a URL request.

Since WMS allows users to access WMS servers through a common interface over the Web using a standard web browser, the format of spatial data behind the WMS is not a concern anymore. Different applications may publish their data with different formats through WMS over the Web. However, WMS allows user to easily integrate and visualize these different format data together. Because WMS allows spontaneous access to different datasets that may be located at different data

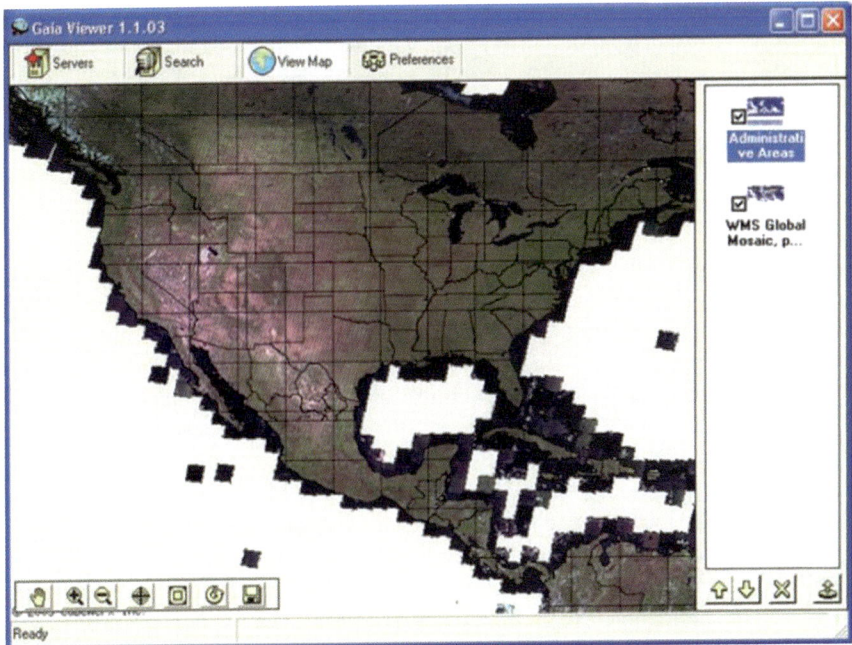

Fig. 1.13 A composite map from two separate web map servers. (Source: Zhang and Li 2005)

servers, it provides a foundation for users to instantly access diverse remote data over the Web.

A composite map can be made by overlaying two or more maps from different URLs. The overlaid two or more maps from different URLs are called "cascaded" layers. A WMS can cascade the content of another WMS by changing the cascaded attributes of the affected layers. A WMS allows the overlaid maps to be visible by supporting transparent backgrounds. Figure 1.13 shows a composite map, which includes two layers—one layer is a pan sharpened Global Mosaic image and the other is a map of Administrative Areas. These two layers' data, in fact, come from two separate web map servers. One map server is OnEarth web Map Server (http:// wms.jpl.nasa.gov/wms.cgi), which supplies the Global Mosaic image layer. The other map server is CubeSERV web map server (http://demo.cubewerx.com/demo/ cubeserv/cubeserv.cgi), which provides the Administrative Areas layer. The lower map layer in Fig. 1.13 is visible through the upper map layer by using a transparent background.

WMS allows users to build customized maps by using different styles. A Style in WMS decides how a layer will be presented in a map. The different styles unlock the graphic definitions of the data, and allow flexible cartographic representations with high quality. For example, users may use blue as a style for displaying river features and use black as another style for drawing road features. Each layer in a WMS may have a number of different styles. Users may request different views of the same data through different WMSs. Users can specify these views by using a Style Language Descriptor (SLD) XML language introduced by OGC. A Styled Layer Descriptor-enabled WMS allows users to map feature data from a WFS using user-defined symbols.

Users may use SLD symbols in a WMS through three ways. First, users may interact with a WMS using *HTTP GET* through a request that refers to a remote SLD. Second, users may also use the *HTTP GET* method but include the *SLD XML* document online with the *GET* request in a *CGI* parameter. Third, users may also interact with a WMS using *HTTP POST* with the *GetMap* request encoded in XML and including an embedded *SLD*. Note that the WMS server has no prior knowledge of the SLD contents. It is the WMS client software that allows users to interactively define how a map appears and to construct the necessary SLD "on-the-fly". The SLD document must be at a web location accessible to the WMS client.

Unlike WFS, which allows users directly to access specific feature data in GML from a WFS sever, WMS does not allow users to access specific data from its server. However, WMS produces maps for users (note that GML is only concerned with the representation of geographic data content and does not specify how data should be presented). WMS can produce dynamic maps based on the WFS queried feature-level results. In addition, WMS itself provides parameters and functions to enable users to query and integrate data from diverse sources. Like WFS, WMS also allows users to query a WMS server about features displayed on a map. In general, WMS provides mechanisms to render maps in a graphical format from a WMS sever; however, it has no capability to retrieve actual feature data or coverage data values.

There are two types of WMS—Basic WMS and Queryable WMS. Basic WMS provides two operation protocols: *GetCapabilities* and *GetMap*. Queryable WMS also supports the operation *GetFeatureInfo*. *GetCapabilities* allows a WMS client to instruct a WMS server to expose its mapping content and processing capabilities and return service-level metadata. *GetMap* enables a WMS client to instruct multiple WMS servers to independently craft "map layers" that have identical spatial reference system, size, scale, and pixel geometry. The WMS client can then display these overlays in a specified order and transparency, such that the information from several sources is rendered for immediate human understanding and use. *GetFeatureInfo* enables a user to use the point coordinates in a raster map to inquire about metadata values of the feature(s) represented there. The *GetFeatureInfo* operation is designed to give users with more information about features in the pictures of maps.

1.4.3 Web Coverage Service

The OGC Web Coverage Service (WCS) promotes geospatial data sharing as "coverages", which represent space/time-varying geospatial phenomena (OGC 07-011 2006). The WCS is based on the GML Application Schema for Coverages (OGC 09-146r2 2012), OWS Common (OGC 06-121r9 2010), and OGC Abstract Topic 6 (OGC 07-011 2006).

Like WFS and WMS, WCS also allows users to retrieve geospatial information or data based on spatial constraints and other query criteria (OGC 09-110r4 2012). Unlike WFS, which retrieves specific geospatial features from a WFS sever, WCS returns coverages as a specialized class of features to represent space/time-varying phenomena. Unlike WMS, which returns static maps or pictures from the server, WCS provides available data together with their detailed descriptions. The WCS defines a rich syntax for users to make requests for the spatial data. It not only portrays the spatial data but also returns data with its original semantics instead of pictures for users to make interpretation, extrapolation, and other spatial analysis (OGC 09-110r4 2012).

WCS uses the coverage model of the OGC GML Application Schema for Coverages (OGC 09-146r2 2012). The term "GML coverage" refers to the concrete data structure defined in the GML Application Schema for Coverages (OGC 09-110r3 2010). It supports all coverage types supported by the OGC WCS Standard version 2.0 and higher versions (OGC 09-110r3 2010). The GML coverage structure includes a *domainSet* describing the coverage's domain, a *rangeSet* describing the range values ("pixels", "voxels") of the coverage, and a *rangeType* defining the coverage's range set data structure (OGC 09-110r3 2010). A *domainSet* is based on *GML:DomainSet*, a *rangeSet* is based on *GML:RangeSet*, and a *rangeType* is based on the *SWE (Sensor Web Enablement) Common DataRecord* (OGC 08-094 2010). Figure 1.14 shows the Coverage Structure. Please note that the coverage model is independent from WCS, and can be used by a wide range of coverage application domains and service types for interoperability.

The coverage types supported by the OGC WCS Standard version 2.0 and higher (OGC 09-110r3 2010) include:

* MultiPointCoverage (ISO 19123: CV_DiscretePointCoverage)
* MultiCurveCoverage (ISO 19123: CV_DiscreteCurveCoverage)

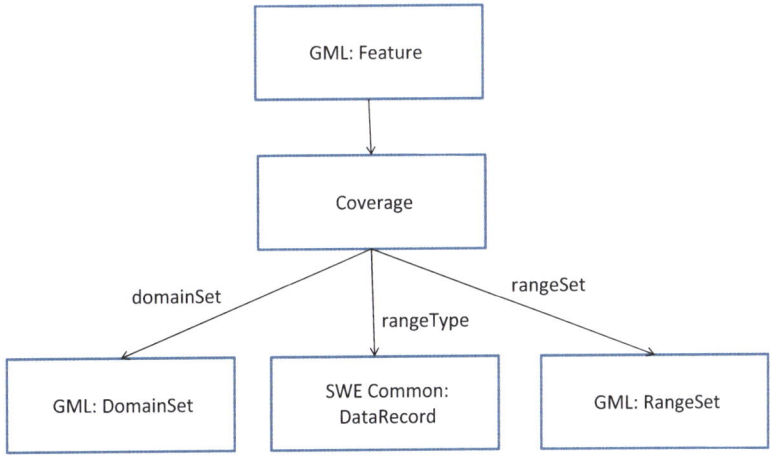

Fig. 1.14 The coverage structure. (Adapted from OGC 09-146r2 2012)

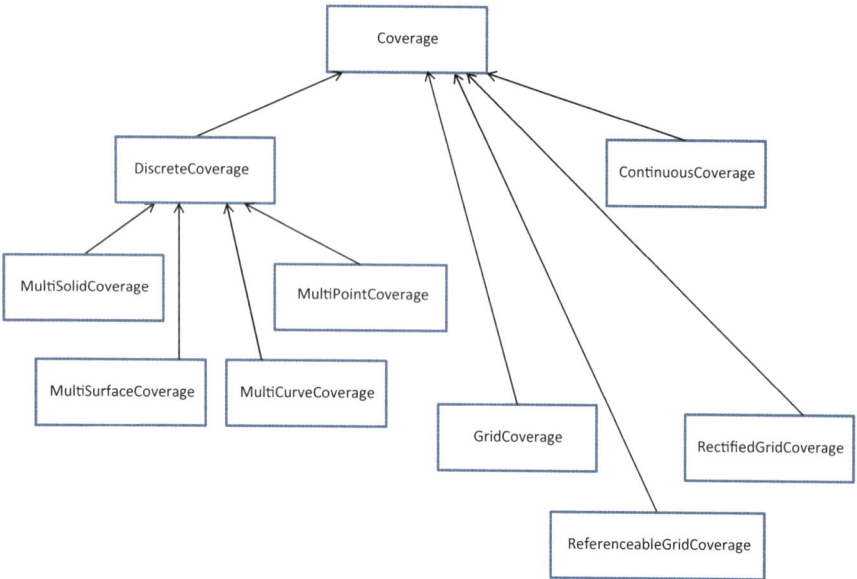

Fig. 1.15 The coverage type hierarchy. (Adapted from OGC 09-146r2 2012)

- MultiSurfaceCoverage (ISO 19123: CV_DiscreteSurfaceCoverage)
- MultiSolidCoverage (ISO 19123: CV_DiscreteSolidCoverage)
- GridCoverage (ISO 19123: CV_DiscreteGridPointCoverage)
- RectifiedGridCoverage (ISO 19123: CV_DiscreteGridPointCoverage)
- ReferenceableGridCoverage (added to GML via Change Request (OGC 07-112r3 2011))

Figure 1.15 shows the Coverage type hierarchy. From this figure, it can be seen that *MultiPointCoverage, MultiCurveCoverage, MultiSurfaceCoverage*, and *MultiSolidCoverage* are discrete coverages, which consist of either spatial or temporal geometry objects, finite in number. A *ContinueCoverage* is a coverage that can give users different values for the same feature attribute at different direct positions in a single spatiotemporal object. A *GridCoverage* is a discrete point coverage in which the domain is a geometric grid of points. A *RectifiedGridCoverage* is a discrete point coverage based on a rectified grid. A *RectifiedGridCoverage* is similar to a *GridCoverage* except that the points of the *RectifiedGridCoverage* are geometrically referenced. A *ReferenceableGridCoverage* has a domain geometry that is a subtype of *GML:ReferenceableGrid*.

WCS provides three operations: *GetCapabilities*, *DescribeCoverage*, and *GetCoverage*. The *GetCapabilities* operation allows users to request information about capabilities and coverages offered by a WCS server. The *DescribeCoverage* operation permits users to request detailed metadata on the selected coverages offered by a server. The *GetCoverage* operation allows users to request a coverage that is comprised of selected range properties at a selected set of spatio-temporal locations and expedited in some coverage encoding format. Normally, users first issue a *Get-Capabilities* request to the WCS server to get an up-to date listing of available data. Based on the listing, users may then issue a *DescribeCoverage* request to find out more detail information about particular coverages offered. Finally, users may issue a *GetCoverage* request to retrieve a coverage or a part of a coverage. Through these three operations, WCS can deliver coverages in a variety of data formats, such as GML, GeoTIFF, NITF (National Imagery Transmission Format Standard), and HDF (Hierarchical Data Format).

WCS requests and responses use the HTTP GET protocol, the HTTP POST protocol, and the SOAP protocol to transfer XML data and binary coverage data.

1.4.4 Web Processing Service

The OGC Web Processing Service (WPS) is based on the OGC WPS Interface Standard, which gives rules for standardizing inputs and outputs (requests and responses) of geospatial processing services (OGC 05-007r7 2007). Geospatial processes refer to any algorithm, calculation or model that operates on spatially referenced data. The OGC WPS Interface Standard also defines how users can request the execution of a process and how the output from the process is handled. The OGC WPS facilitates the publishing of geospatial processes and helps users to discover and bind these processes by making machine-readable binding information and human-readable metadata available.

Any sort of GIS functions such as pre-programmed calculation or computation models that operate on spatially referenced data can be converted to a WPS. A WPS may offer simple calculations such as subtracting one set of spatially referenced numbers from another, or complicated spatial model functions such as a land use/cover change model. The data required by the WPS must be available through a network or the Web from a server. WPS has mechanisms to identify the geospatial data required by the decision models, execute the calculation of the models, and manage the output from the calculation, so that they can be accessed by the users. Both vector and raster data can be processed by WPS. The data can include image data formats such as GeoTIFF, or data exchange standards such as GML or Geolinked Data Access Service (GDAS).

The design purpose of the WPS is to allow users to execute the spatial processes/models without knowledge of the underlying physical process interface or API (Application Programming interface). Because WPS uses a generic interface, users can wrap it with other existing OGC services for providing geospatial processing functions. WPS facilitates the use of geospatial processing functions over the Web by moving the GIS functionality to the Internet. The successful use of WPS may greatly reduce the amount of programming required for implementation of geospatial models.

Three operations are required for a WPS interface (OGC 05-007r7 2007): *GetCapabilities*, *DescribeProcess*, and *Execute* operations. The *GetCapabilities* operation allows users to request and receive back service metadata (or capabilities) documents that describe the abilities of the implemented decision models, such as names and general descriptions of the processes provided by a WPS. The *DescribeProcess* operation grants users the right to request and receive back detailed information about one or more spatial model(s) that can be executed, including the input parameters and formats and the outputs. The *Execute* operation allows users to run a specified spatial model implemented by the processing services, which uses the provided input parameter values and returns the produced outputs. These three operations have many similarities to other OGC Web Services such as WFS, WMS, and WCS.

Because WPS is a generic interface, it does not specify any specific processes that are supported. However, when users implement each WPS they must specify the processes that it supports and the associated inputs and outputs of the supported processes. To achieve interoperability, service providers must specify the specific implemented spatial models in a separate document called an Application Profile. An Application Profile includes two required elements and two optional elements. The two required elements are: a) an OGC URN (Uniform Resource Name) that uniquely identifies the process, and b) a reference response to a *DescribeProcess* request for that process. The two optional elements are: a) a human-readable document that describes the process and its implementation, and b) A WSDL (Web Services Description Language) description for that process. WPS Application Profiles are used by web service registries for search metadata for multiple service instances.

WPS is compatible with both WSDL and SOAP (Simple Object Access protocol), which are two protocols for the implementation of Web Services in computer network. SOAP can be used to package WPS requests and responses. The use of SOAP to bind WPS requests offers the capability to add security certificates and encryption to web-based geoprocessing transactions (OGC 05-007r7 2007). WPS supports the use of WSDL for an individual WPS process as well as the WPS that may include several processes. Like WMS and WFS, WPS also makes requests through *HPPT Get* and *Post*. WPS uses two different methods for the provision of input data: (1) data can be embedded in the Execute request, or (2) referenced as a web accessible resource.

A WPS process is normally an atomic function that performs a specific geospatial calculation or modelling. An application may need to chain several WPS processes into repeatable workflows to achieve its goal. Three methods can be used for chaining WPS processes (OGC 05-007r7 2007). The first method is to use a BPEL (Business Process Execution Language) engine to orchestrate a service chain that has one or more WPS processes. The second method is to design a WPS process to call a sequence of web services including other WPS processes. The third method is to encode simple service chains as part of the *Execute* query via the GET interface.

In general, WPS provides a great advantage to promote geospatial process interoperability and share geospatial process functions or models over the Web. WPS allows service developers to reuse significant amounts of code in the development of their web applications, and it also facilitates ease of understanding among the web application developers.

1.4.5 Catalogue Service and Services-Oriented Architecture

The OGC Catalogue Service (CS) provides a standard interface to support discovery, access, maintenance, and organization of catalogues of geospatial information and related resources (OGC 07-006r1 2007). The OGC Catalogue Service provides catalogues for the above introduced OGC data and processing services, and supports the ability to publish and search collections of descriptive information (metadata) for data, services, and related information objects in multiple distributed computing environments, including the Web environment. The essential purpose of a catalogue service is to enable users to locate, access, and make use of resources in an open, distributed system by providing facilities for retrieving, storing, and managing many kinds of resource descriptions. Figure 1.16 illustrates one example of registering a WFS over the Web via a catalogue service.

A catalogue service has two main functions—discovery and publication. Discovery means that users seek to find resources of interest through simple browsing or by sophisticated query-driven discovery that specifies simple or advanced search criteria. The catalogue performs the search and returns a result set which contains all registry objects that satisfy the search criteria. Figure 1.17 shows one example of searching (discovering) web services by category "transportation" via a catalogue service. OGC Catalogue Service supports distributed searches. When decision-makers perform a distributed search, the requested message is forwarded to one or more affiliated catalogues to enlarge the total search space.

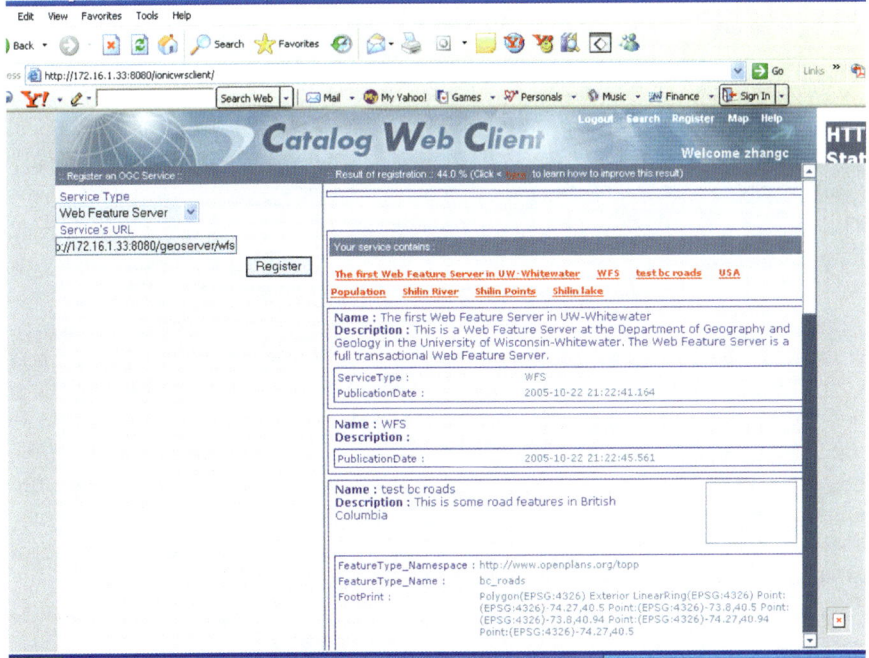

Fig. 1.16 An example of registering a WFS over the Web via a catalogue service

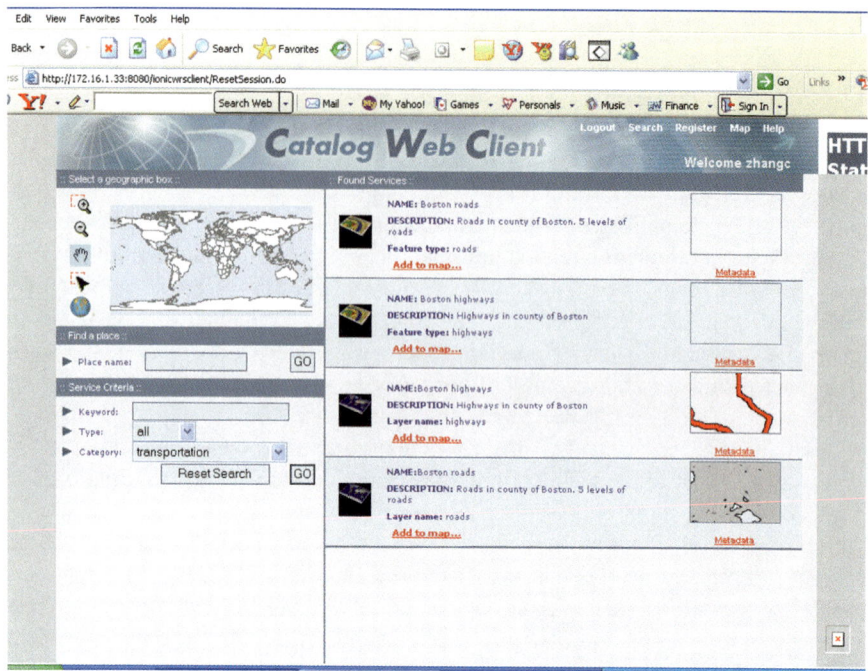

Fig. 1.17 An example of searching (discovering) web services by category "transportation" via a catalogue service

OGC Catalogue Service allows web services to publish resources in two ways: the harvest operation and the transaction operation. The harvest operation is the basic publication in which the Catalogue Service attempts to harvest a resource from a specified network location, thereby realizing a 'pull' model for publishing registry objects. If the Catalogue Service successfully retrieves the resource and can handle it, then one or more corresponding registry objects are created or updated. The second way of publication is the transaction operation, which is an enhanced publication. The transaction operation supports the modification of the Catalogue Service content through a public interface that allows a 'push' style of publication. A user may insert, update, or delete registry entries according to criteria specified in the requested message. Typically this operation is subject to some kind of access controls such that only authorized users may perform such actions. Users may adopt both operations at different stages. At the initial stage of service registry, users may use the transaction operation, which allows service providers more control in their initial registry. After the service has been registered, users may use the harvest operation to minimize human interference in the daily maintenance.

There are four mandatorily required operations and three optional operations for implementation of OGC Catalogue Services:

- *GetCapabilities* (required)
- *DescribeRecord* (required)
- *GetRecords* (required)

- *GetRecordsById* (required)
- *GetDomain* (optional)
- *Havest* (optional)
- *Transaction* (optional)

The *GetCapabilities* operation allows users to retrieve service metadata describing catalogue service instances. Like WFS, WMS, and WCS, the response to a *GetCapabilities* operation should be an XML document containing service metadata about the server. The *DescribeRecord* operation allows users to find out elements of the information model supported by the catalogue service. The response to the operation can be part or the entirety of the information model to be described. The *GetRecords* operation allows users to search and present records. The operation allows users to obtain information about the found records, such as the number of records found and/or a reference to the type or schema of the records found. It's not necessary for the catalogue server to process the *GetRecords* operation immediately. The server may process the operation at a later time, taking as much time as is required to complete the operation. Once the operation is completed, a response message or an exception message may be sent to the server. The *GetRecordById* operation allows users to retrieve the default representation of catalogue records using their identifiers. This operation assumes that a previous query such as a *GetRecords* operation has been performed to obtain the identifiers that may be used with this operation. This operation is included as a convenient short form for retrieving and linking to records in a catalogue.

GetDomain, *Transaction*, and *Havest* are three optional operations. The *GetDomain* operation allows users to get runtime information about the range of values of a metadata record element or request parameter. The *GetDomain* operation tries to generate useful information about a specific request parameter or property if it can. However, it may be possible that a catalogue is unable to determine anything about the values of a property or request parameter beyond the basic type. Under this case a type reference or a type description will be returned. The *Transaction* operation allows users to create, modify, and delete catalogue records. The operation may be used to "push" data into the catalogue. The *Harvest* operation also can be used to create or update records in the catalogue. However, the *Harvest* operation is different from the *Transaction* operation. The *Harvest* operation only references the data to be inserted, updated, or deleted in the catalogue, and it relies on the catalogue service to resolve the reference, fetch that data, and process it into the catalogue. Thus, the *Harvest* operation can be considered as an operation that "pulls" data into the catalogue. The *Harvest* operation has two modes of operation: *synchronous* mode and *asynchronous* mode. In the *synchronous* mode, the *Harvest* operation processes the user's request immediately and sends the results back to the user while the user waits. In the *asynchronous* mode, however, the *Harvest* operation only sends the user an immediate acknowledgement that the request has been successfully received. The server may process the request at any time it likes, and then sends back the results to a URI specified in the original *Harvest* request. In the *asynchronous* mode, *Harvest* requests can be run for a period of time longer than that most HTTP time-outs will allow.

Fig. 1.18 Three elements in a services-oriented architecture

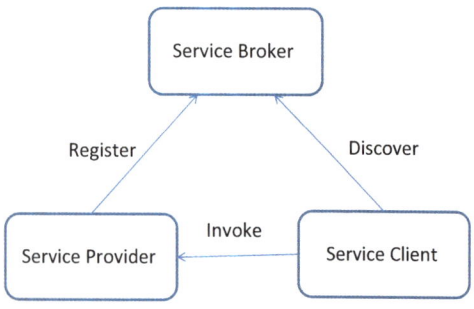

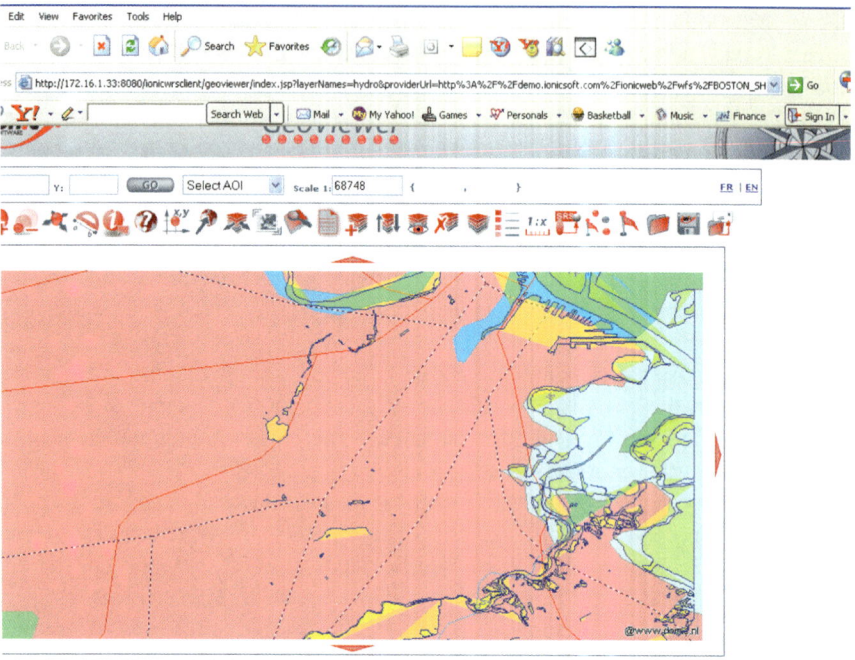

Fig. 1.19 Integrate different formats of data from disparate servers by invoking the WFS and WMS services (*bind*)

The architecture that makes full use of web catalogue services and other data web services, such as OGC WFS, WMS, WCS, and WPS, is called Services-Oriented Architecture (SOA). A pure architectural definition of SOA might be "an application architecture within which all functions are defined as independent services with well-defined invokable interfaces which can be called in defined sequences to form business processes" (Kishore et al. 2003). In another word, the SOA can be defined as a system in which resources are made available to other participants in the network as independent services that are accessed in a standardized way.

There are three elements in a Services-Oriented Architecture (SOA)—service provider, service broker, and service client. Figure 1.18 shows the three elements in a SOA. Service provider supplies heterogeneous geoprocessing functions (or models) and geospatial data from disparate sources via OGC Web Services such as WPS, WFS, WMS, and WCS. Service broker provides a registry for available services. Service broker may use OGC Catalogue Services (CSs) to register and manage the data and processing services and allow users to search for these services. Service clients search contents of catalogue services to find the services of interest. Service clients may also combine individual data and processing services found through catalogue web services to carry out their application tasks. In the SOA, the three components—service provider, service client and service broker work together. A service provider publishes services to a service broker. A service client finds required services using a service broker and binds to them. The binding from the service client to the service provider should loosely couple the services. This means that the service requester has no knowledge of the technical details of the provider's implementation, such as the programming language, deployment platform, and so forth. Figure 1.19 shows a map that integrates different formats of data from disparate servers by invoking the WFS and WMS services (i.e., bind).

The SOA aims to create and reconfigure an application to support new and rapidly changing situations. It moves away from monolithic systems towards distributive systems with interoperable components. With SOA, an application can be built without a priori dependency on other applications. The web services in an application can be added, modified, or replaced without impacting other applications. Thus, the SOA provides more flexible loose coupling of resources than do traditional system architectures. With SOA, applications can access web services through the web without the concern that how each service is implemented.

1.5 Chapter Summary

This chapter introduces the background information about geospatial data interoperability and the state-of-the-art technologies for achieving geospatial data interoperability, such as GML, SVG, and Geospatial web services. Although many GIS databases have been developed, geospatial data interoperability is still a challenge faced to geospatial community. GML as a standard data exchange format aims to achieve the goal of data interoperability by providing mechanisms for data sharing and reuse at the feature level over the Web. However, GML has been designed to uphold the principle of separating content from presentation. Thus SVG can be used to style GML data for presentation. As a vector graphic, SVG can display high quality maps. While GML provides a means to encode and transport geospatial features into XML, SVG provides a means to display these GML-coded geospatial features into vector maps on the Web. One issue of concern is how to conduct queries and extract features from the database to respond to user requests. The Geospatial Web Services Implementation Specifications developed by the OGC serve this role. Specifically, (1) Web Feature Service (WFS) allows a client to retrieve, query, and manipulate feature-level geospatial data encoded in GML (Geography Markup Language) from multiple sources; (2) Web Map Service is capable of creating and displaying maps that come simultaneously from multiple heterogeneous sources in a standard image format; (3) Web Coverage Service provides access to potentially detailed and rich sets of geospatial information in forms that are useful for client-side rendering, multi-valued coverage, and input into scientific models and other clients; (4) Web Processing Service defines rules for standardizing inputs and outputs (requests and responses) of geospatial processing services; (5) Catalogue Service provides catalogues for OGC web services and supports the ability to publish and search collections of descriptive information (metadata) for data, services, and related information objects. The architecture that makes full use of web catalogue services and other web services, such as OGC WFS, WMS, WCS, and WPS, is called Services-Oriented Architecture (SOA). The SOA moves away from monolithic systems towards distributive systems with interoperable components, and implementations of the SOA may decrease problems in data and model duplication and maintenance.

References

Anderson G, Moreno SR (2003) Building web based spatial information solutions around open specifications and open source software. Trans GIS 7:447–466

Apache (2006) Xalan-Java Version 2.7.1. http://xml.apache.org/xalan-j/. Accessed 16 Jan 2014

Arctur D, Hair D, Timson G et al (1998) Issues and prospects for the next generation of the spatial data transfer standard (SDTS). Int J Geogr Info Sci 12:403–425

Badros, GJ et al (2001) A constraint extension to scalable vector graphics. In: Proceedings of the 10th international conference on World Wide Web, Hong Kong, China, April 2001, pp 489–498

Cömert C (2004) Web services and national spatial data infrastructure (NSDI). In: Proceedings of geo-imagery bridging continents, XXth ISPRS Congress, Istanbul, Turkey, Commission 4, 12–23 July 2004. http://www.isprs.org/istanbul2004/comm4/papers/ 365.pdf. Accessed 4 Sept 2005

Choicki J (1999) Constraint-based interoperability of spatiotemporal databases. Geoinformatica 3:211–243

Devogele T, Parent C, Spaccapietra S (1998) On spatial database integration. Int J Geogr Info Sci 12:335–352

Jia W et al (2004) Web service based web feature service. In: Proceedings of geo-imagery bridging continents, XXth ISPRS Congress, Istanbul, Turkey, Commission 4, 12–23 July. http://www.isprs.org/istanbul2004/comm4/papers/334.pdf. Accessed 4 Sept 2005

Kay MH (2013) SAXON: The XSLT and XQuery Processor. http://saxon.sourceforge.net/. Accessed 30 Oct 2014

Kishore C et al (2003) Migrating to a service-oriented architecture. http://www-128.ibm.com/developerworks/library/wsmigratesoa/. Accessed 30 Oct 2014

NIST (National Institute of Standard and Technology) (1994) Federal information processing standard publication 173-1, spatial data transfer standard. US Department of Commerce, Washington DC

Noronha V (2000) Towards ITS map database interoperability—database error and rectification. GeoInformatica 5:345–373

OGC White Paper (2001). Introduction to OGC web services. http://www.opengeospatial.org/pressroom/papers. Accessed 16 Jan 2014

OGC document 04-094 (2005) Web feature service implementation specification, version 1.1.0. http://www.opengeospatial.org/specs/?page=specs. Accessed 4 Sept 2005

OGC 07-011 (2006) The OpenGIS® abstract specification topic 6: schema for coverage geometry and functions. http://www.opengeospatial.org/standards/as. Accessed 16 Jan 2014

OGC 06-042 (2006) OpenGIS® web map server implementation specification. http://www.opengeospatial.org/standards/wms. Accessed 16 Jan 2014

OGC 05-007r7 (2007) OpenGIS® web processing service. http://www.opengeospatial.org/standards/wps. Accessed 16 Jan 2014

OGC 07-036 (2007) OpenGIS® geography markup language (GML) encoding standard, version 3.2.1. http://www.opengeospatial.org/standards/gml. Accessed 16 Jan 2014

OGC 07-006r1 (2007) OpenGIS® catalogue services specification. http://www.opengeospatial.org/standards/cat. Accessed 16 Jan 2014

OGC 06-121r9 (2010) OGC web service common specification, version 2.0. http://www.opengeospatial.org/standards/common. Accessed 16 Jan 2014

OGC 09-110r3 (2010) Web coverage service (WCS) core interface standard, version 2.0. http://www.opengeospatial.org/standards/wcs. Accessed 16 Jan 2014

OGC 09-025r1 (2010) Open GIS web feature service 2.0 Interface standard. http://www.opengeospatial.org/standards/wfs. Accessed 16 Jan 2014

OGC 08-094 (2010) OGC® SWE common data model encoding standard. http://www.opengeospatial.org/standards/swecommon. Accessed 16 Jan 2014

OGC 10-100r3 (2011) Geography markup language (GML) simple features profile (with Corrigendum). http://www.opengeospatial.org/standards/gml. Accessed 16 Jan 2014

OGC 07-112r3 (2011) Add implementation of ISO 19123 CV_ReferenceableGrid to GML. http://www.opengeospatial.org/standards/cr. Accessed 16 Jan 2014

OGC 09-146r2 (2012) OGC® GML application schema—coverages. http://www.opengeospatial.org/standards/wcs. Accessed 16 Jan 2014

OGC 09-110r4 (2012) OGC® WCS 2.0 interface standard—core: Corrigendum. http://www.opengeospatial.org/standards/wcs. Accessed 16 Jan 2014

OGC 10-129r1 (2012) OGC® Geography markup language (GML)—extended schemas and encoding rules. http://www.opengeospatial.org/standards/gml. Accessed 16 Jan 2014

Peng ZR, Tsou MS (2003) Internet GIS: distributed geographic information services for the internet and wireless networks. Wiley, New York

Peng ZR, Zhang C (2004) The roles of geography markup language, scalable vector graphics, and web feature service specifications in the development of internet geographic information systems. J Geogr Syst 6:95–116

Siki Z (1999) GIS data exchange problems, solutions. Periodica polytechnika. http://bme-geod.agt.bme.hu/staff_h/siki/gisexch.htm. Accessed 16 Jan 2014

Stephan E et al (1993) Virtual data set: an approach for the integration of incompatible data. In: Proceedings of Auto Carto 11. American congress on surveying and mapping, Bethesda, pp 93–102

W3C XML-1 (2004) XML schema part 1: structures. http://www.w3.org/TR/xmlschema-1. Accessed 16 Jan 2014

W3C XML-2 (2004) XML schema part 2: datatypes. http://www.w3.org/TR/xmlschema-2. Accessed 16 Jan 2014

W3C (2001) Scalable vector graphics (SVG) 1.0 specification. http://www.tka4.org/materials/lib/Code/FileFormats/Vector2D-SVG/REC-SVG-20010904.pdf. Accessed 16 Jan 2014

Zhang C, Li W (2005) The roles of web feature and web map services in real time geospatial data sharing for time-critical applications. Carto Geogr Info Sci 32:269–283

Zhang C, Li W, Peng ZR et al (2003) GML-based interoperable geographical database. Cartography 32:1–16

Chapter 2
Conceptual Frameworks of Geospatial Semantic Web

2.1 Semantic Problem of Spatial Data

With Web Services and Service-Oriented Architecture, users can find, share, and re-use geospatial information more easily over the Web than before. In fact, to facilitate exchange and sharing of spatial data by building on initial expenditures, Spatial Data Infrastructures (SDIs) have been developed in many countries in the past two decades (e.g. Crompvoets et al. 2004; Masser 2005; Rajabifard et al. 2006). The SDIs based on the open standards and OGC web service technologies offer the potential to overcome the heterogeneous problems of legacy GIS databases and facilitate sharing geospatial data in a cost effective way (Askew et al. 2005; Peng and Zhang 2004a, b; Zhang et al. 2003; Zhang and Li 2005; Tait 2005). Many local, regional, and global SDIs have been developed in the world based on the open standards and OGC web services (e.g. Williamson et al. 2003; Crompvoets et al. 2004; Tait 2005; Mansourian et al. 2006). Although these SDIs have undoubtedly improved sharing and synchronization of geospatial information across the diverse resources, there are limitations in the currently implemented SDIs. It can still be difficult to find data sources from the currently implemented SDIs for the following reasons:

Firstly, the currently implemented SDIs only emphasize technical data interoperability and cannot resolve semantic heterogeneity problems in spatial data sharing (Wiegand and García 2007). Differences in the semantics used in diverse data sources are one of the major problems in spatial data sharing and data interoperability (Bishr 1998). Secondly, with the currently implemented SDIs, it is only possible to search and access geospatial data and services by keywords in metadata and it is impossible to directly search and access geospatial data and services based on their contents (Farrugia and Egenhofer 2002). This causes a problem for novice portal users who may not know which keywords to use or may not even know they should try many keywords (Wiegand and García 2007). In addition, a keyword-based search may have a low recall if a different terminology is used and/or a low precision if terms are homonymous (Lutz 2007). On the other hand, the keyword search may sometimes bring an overwhelming number of search results. As a result, users may have to spend a lot of time sifting through undesirable query results

© Springer International Publishing Switzerland 2015
C. Zhang et al., *Geospatial Semantic Web*, DOI 10.1007/978-3-319-17801-1_2

before finding the desired data set (Wiegand and García 2007). Thirdly, although SDIs aim to make discovery and access to the distributed geographic data more efficient, the catalogue services currently used in SDIs for discovering geographic data do not allow expressive queries and also do not take into account more than one data source that might be required to answer a question (Lutz and Kolas 2007). It is not possible to automatically discover several data sources that only in combination can provide the information required to answer a given question. However, it is unrealistic to expect that one web service or one data source can fulfill exactly the needs of a user's request. Therefore, SDIs need a semantic-based approach that can reason about a service's capability to a level of detail that permits their automatic discovery and composition.

In fact, the semantic problem in spatial data sharing and data interoperability has been recognized in the literature for a long time (Bishr 1998). Bishr (1998) classified three types of heterogeneity in GIS databases: (1) Semantic heterogeneity, which is caused by different descriptions in the underlying databases to comply with various disciplines; (2) Schematic heterogeneity, which refers to the facts that objects in one database may be considered as properties in another database or object classes may have different aggregation or generalization hierarchies; (3) Syntactic heterogeneity, which is caused by different implementation models or different geometric representations of geographic objects, such as raster and vector representations. OGC web services and GML data can help to resolve schematic and syntactic heterogeneity problems to achieve data interoperability at the schematic and syntactic levels. However, OGC web services and GML data cannot help to resolve the semantic heterogeneity problem. The semantic heterogeneity problem is usually the source of most of data sharing problems.

The semantic heterogeneity refers to disagreements about meaning, interpretation or intended use of the same or related data. It may cause problems in meaningful data sharing. For example, data exchanged may not be correctly interpreted and used if there is a semantic problem. Although the definition of the semantic heterogeneity is not totally clear, it is well known in literature that the semantic heterogeneity is an important impediment to spatial data sharing and data interoperability. Different GIS databases may use different descriptions or words to model or term their spatial data. For example, one GIS database may call a real world spatial point feature as a "house" while another GIS database may call the same feature as a "building". For another example, the same word "street" may have different meanings in different databases developed by different agencies. In one database, the "street" may refer to the small roads in residential areas while it may refer to the highway in other areas in another database. This semantic heterogeneity causes problems for transparent communication at the semantic level. Computers may not understand the meanings of the data appropriately thus cannot automatically aggregate different spatial data over the Web.

2.2   What is Geospatial Semantic Web? Why Do We Need Geospatial Semantic Web?

Semantic Web was recently proposed to overcome the semantic heterogeneity problem and provide computers meaningful web contents (Berners-Lee et al. 2001). Most of the web's contents today are designed for human to read, not for software programs to process the semantics meaningfully. Semantic Web aims to bring structures to the meaningful contents of Web pages and allow software programs to process and "understand" the data on the Web pages. Via Semantic Web, software programs are able to use structured collections of information and sets of inference rules to conduct automated reasoning.

Geospatial Semantic Web is an extension of the current Web, where geospatial information is given well-defined meaning by ontology and thus geospatial contents can be discovered, queried, and consumed automatically by software (Zhang et al. 2007). Geospatial Semantic Web aims to add computer-processable meaning (semantics) to the geospatial information on the World Wide Web. Because there are different encodings of geospatial semantics in GIS databases, it is a challenge to automatically process requests for geospatial information over the Web. The Geospatial Semantic Web concept was proposed to address the vexing semantic challenge and achieve automation in service discovery and execution (Peng and Zhang 2005). As aforementioned, while GML and OGC web services provide syntactic ways to encode geospatial information over the Web, they are unable to capture the semantic information of GIS data. Thus a Web user still has difficulty to find an appropriate spatial data set for a specific task using one of the current search engines, because geospatial data sets encoded using GML and OGC web services are lack of semantic information and software programs are unable to understand the meanings expressed by geospatial data contents and requests. However, Geospatial Semantic Web is capable of capturing, analyzing, and tailoring geospatial information beyond the purely lexical and syntactic level (Egenhofer 2002).

Although OGC web services have undoubtedly improved sharing and synchronization of geospatial information across diverse resources, there are limitations in the currently implemented OGC web services (Zhang et al. 2010a):

First, although the OGC web services facilitate data interoperability at the syntactical level via standard interfaces, they cannot resolve data interoperability problems at the semantic level. One of the major problems in spatial data sharing and data interoperability is the semantic heterogeneity of spatial data. The OGC web service descriptions only allow the specification of the syntax of basic service contents, such as *operation metadata*, *feature type lists*, and *filter capabilities*, and they provide no semantic descriptions of the meaning of these contents. Two identical XML descriptions may mean very different things depending on the contexts of their uses. In addition, the WFS specification of the outputs of each call to the services similarly lacks semantic definitions. All defined search operations return results using the same data structure, regardless of what information is requested. For example, a *Building* feature contains a field *Commercial Building,* which is used to describe buildings in commercial areas, and a field *Residential Building,* which is used to describe houses in residential areas. Even if the types of *buildings* specified in a *Building* file were clearly identified in a *Type* field by the interface designer, the OGC WFS description provides no uniform way of enabling such interpretations. It is up to the WFS clients to recognize the values in these fields, which indicate whether it is a commercial or a residential building.

Second, the OGC web services only make it possible to search and access geospatial data by keywords in metadata, and they do not support content-based search at the semantic level. Because the OGC web service descriptions do not support the semantic specification of service contents and operations, they only allow the semi-structured keyword search based on the metadata. In addition, metadata also have the semantic heterogeneity problem. Different metadata creators may use different names for the same feature. For example, by typing keywords "*school*" and "*Storrs, CT*" in a data system implemented using OGC web services in Connecticut, users may get query results of a bunch of feature-level school data such as *Mansfield Middle School* and *E.O Smith High School* data services for Storrs, Connecticut if their metadata contain exactly these keywords. However, if they use different names for the same feature, it is unlikely that a software program could read and utilize these data services without human assistance. For example, all *FeatureTypes* in a *School* WFS description may be typed as strings, such as *Storrs:SchoolName*, *Storrs:SchoolType*, *Storrs:SchoolID*, *Storrs:SchoolAdministration*, and *Storrs:SchoolSize*. Users may understand the meanings of these strings through additional data dictionaries or experiences and knowledge. However, it is difficult for software programs to automatically infer the meanings of these strings. Software programs cannot understand that *Storrs:SchoolName* refers to *SchoolName* for school data in Storrs, Connecticut, USA. Therefore, with OGC web services, it is difficult to perform intelligent content-based search and users cannot correctly utilize the discovered web services without additional human assistance or programming. Further, metadata contains only limited information to allow users to search. Despite efforts that the geospatial community has put on providing better tools to manage geospatial metadata, the content-based search at the semantic level remains a challenge problem.

Third, without a formal semantic description of OGC web services, it is difficult to allow users and applications to discover, deploy, compose, and synthesize OGC web services automatically. The lack of an explicit semantic in OGC web service descriptions proves to be a major limitation for an automatic capability match. It is unrealistic to expect that advertisements and requests of OGC web services are equivalent, or that there exists a service that can fulfill exactly the needs of the requester. For example, an OGC WFS may be advertised as a *University* data provider, while a requester may need a *College* data service. In order to make OGC web services more practically searchable and ubiquitously available, we need a semantic-based approach to reason about a service's capability. The OGC web services' lack of semantic descriptions of their operations makes it impossible to develop web service clients that can, without human assistance, dynamically find and invoke OGC web services to integrate the semantically heterogeneous geospatial data together.

Geospatial Semantic Web is capable of overcoming the aforementioned problems of OGC web services. The Geospatial Semantic Web technologies, such as Ontology, Description Logic (DL) reasoner, and inference rules, allow OGC web services share and integrate geospatial data contents at the semantic level. Therefore, the systems built on these Geospatial Semantic Web technologies can automatically search and access geospatial data by their contents rather than just by keywords in metadata.

The following section introduces a Geospatial Semantic Web architecture.

2.3   A Geospatial Semantic Web Architecture

Figure 2.1 illustrates a Geospatial Semantic Web architecture for geospatial data sharing. For instant remote data access and exchange, the ontology-based web services are used to access and manipulate geospatial data over the Web from heterogeneous databases. The architecture is based on Service-Oriented Architecture (SOA) and is essentially a collection of ontology-based OGC web services, which communicate with each other by simple data passing or coordinating some activities. It has four major elements: service provider, service broker, service client, and ontology server. The service provider supplies the ontology-based geospatial data;

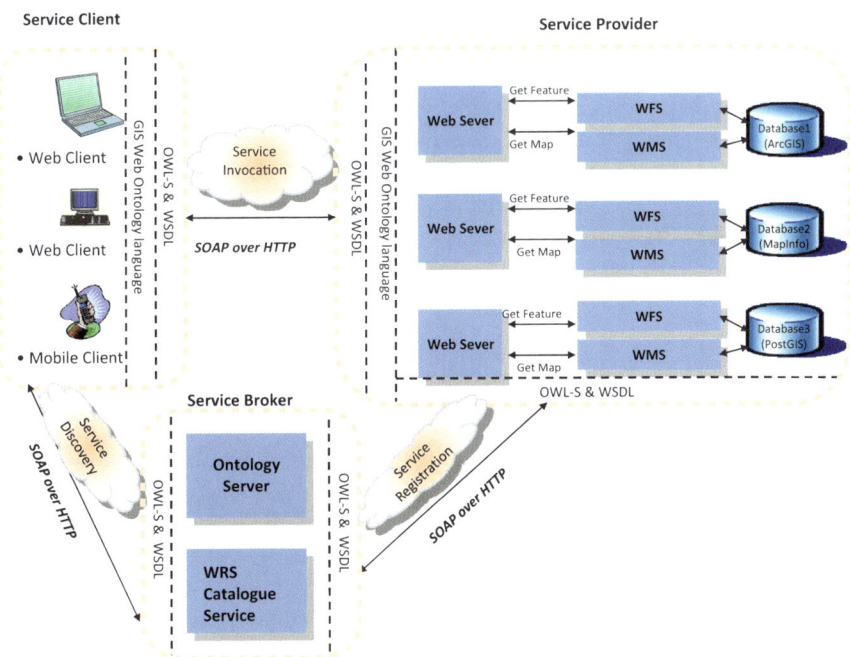

Fig. 2.1 A geospatial semantic web architecture. (Source: from Zhang et al. 2007)

the service client searches and integrates the ontology-based geospatial data from the service providers; and the service broker provides a registry for the available ontology-based web services. The ontology server ensures semantic interoperability of ontologies from the service providers and clients.

The service providers use the ontology-based OGC data services such as Web Feature Service (WFS) and Web Map Service (WMS) to distribute geospatial data connected to a legacy GIS. The ontology-based OGC web services provide a basis to share spatial data at the semantic level from different sources without data conversation. The service broker uses the ontology-based OGC Catalogue Service (CS) to register and manage the data services and allow users to search for these ontology-based data services. Service clients search contents of the ontology-based catalogue services to find the datasets and services of interest, and they can also combine the data services found through the ontology-based catalogue web services. Unlike traditional web services, the ontology-based service clients and providers must maintain their local ontology at both the client side and the provider side to ensure the semantic interoperability. Local ontology at the service client side refers to semantics used by the client users or client applications. Local ontology at the service provider side refers to semantics used by the data providers. These local ontologies may address geospatial relations such as topological relations (e.g. connectivity, adjacency, and intersection among geospatial objects), cardinal directions (e.g. east, west, northwest, and southeast), and proximity relations (e.g. geographical distances among objects). The client ontology must be able to communicate with the provider ontology. Moreover, a service client may need to access multiple service providers to complete a task. Therefore, it is necessary to create mappings of equivalent or related classes and properties in the local ontologies. The ontology server is used to realize this function, and it keeps a taxonomy of geospatial terminologies and maintains consistency for different local ontologies. The service broker uses the ontology server to map the standard OGC catalogue services to the ontology-based catalogue services.

The ontology-based web services are connected via OWL-S (Semantic Markup for Web Services) and Web Service Description Language (WSDL) among the service provider, the service broker, and the service client. OWL-S (http://www.w3.org/Submission/OWL-S/) is an ontology of services built on the Web Ontology Language (OWL) to facilitate automatic web service discovery, invocation, composition, and interoperation. OWL-S allows specification of services in terms of their inputs, outputs, conditions, which have to be true before the service execution (referred to "preconditions" in OWL-S terms), and post-conditions, which represent the state of the environment after the service execution (referred to "effect" in OWL-S terms). OWL-S uses a process model to describe web services. The process model contains a number of atomic processes that can be invoked individually or combined together. The data used by the processes are based on description logic types. Since the standard OGC web services are based on WSDL and geospatial web services use WFS/WMS as extensions of WSDL, rules have been implemented to translate the OWL-S processes from/to WSDL operations and the input/output values of the processes from/to WSDL messages. The Simple Object Access

Protocol (SOAP) binding over HTTP is employed for communication between web services via the Internet. The SOAP essentially provides the envelope for sending web service messages.

Overall, the semantic SOA ensures data interoperability through semantic web services, which offer basic conditions for interoperability by using a standard exchange mechanism among diverse spatial data sources connected over the Internet. The semantic web services provide the interoperable capability of cross-platform and cross-language functionality in the distributed Internet environment. The main advantages of this architecture are its abilities

1) to guarantee data interoperability at semantic level for geospatial data sharing over the Web; and
2) to search and access geospatial web services by content rather than just by keywords in metadata.

Applications and organizations employing this approach can deploy spatial data with different semantics over the Internet so that information from diverse sources with incompatible data formats and semantics can work together transparently across the Internet. The following sections introduce the main technologies used in the architecture, such as ontology, semantic descriptions of geographic information using ontology, the ontology-based catalogue service, and web service composition.

2.3.1   Ontology

The term *Ontology* originated in philosophy and has been used in many different ways. In computer science and information science, an *Ontology* formally represents knowledge using a set of concepts within a domain (Gruber 1993). These concepts can be denoted by a shared vocabulary for their types, properties and interrelationships. In Semantic Web, ontologies are used for organizing information and representing knowledge about the world or some parts of it. There are three types of ontologies: *Domain ontology*, *Upper ontology*, and *Hybrid ontology*. *Domain Ontology* models a specific domain and represents concepts in a specific field. For example, geospatial ontologies belong to the domain ontology in the geospatial field. *Upper Ontology* is a model of the common objects that can be used across a wide range of domain ontologies. It uses a core glossary, which comprises of terms and associated object descriptions as used in various relevant domain sets, to represent knowledge. For example, *Things* is an upper ontology and can be used by different fields. *Hybrid Ontology* is a combination of *Upper Ontology* and *Domain Ontology*. For example, *Geospatial Things* may be considered as a hybrid ontology because it is the combination of an upper ontology *Things* and a domain ontology *Geospatial Feature*.

An ontology model is made up of several elements: classes/concepts; properties and attributes for these concepts; constraints on properties and attributes; relationships between and among class concepts as well as between and among class instances; and instances of concepts. *Class* in ontology is a concept in the domain.

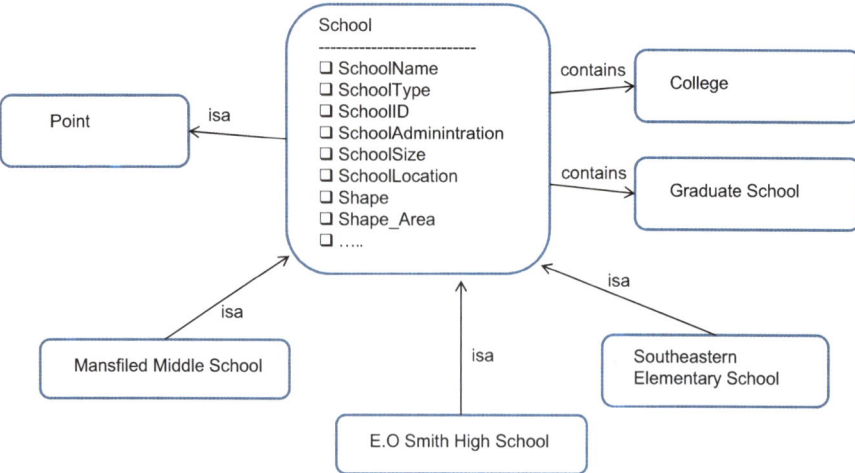

Fig. 2.2 An example of *School* ontology

For example, *School* is a class of an ontology *School*. The *School* class may have a sub-class called *College* and another sub-class called *Graduate School*. Please note that ontology is made up of *Classes* in a formal hierarchy. Ontology may include primitive concepts or foundational concepts from which other concepts can be constructed.

Figure 2.2 illustrates a very simple example of *School* ontology. In this example, *School* is a class with several properties such as *SchoolName*, *SchoolType*, *SchoolID*, *SchoolSize*, and *SchoolLocation*. *School* class also contains several sub-classes such as *College* and *Graduate School*. *School* class has several class instances such as *Mansfiled Middle School*, *E.O Smith High School*, and *Southeastern Elementary School*. In addition, *School* class belongs to *Point* geometry class.

The overall goal of ontology is to be in forms that express a common understanding of the structure of information suitable among applications and software agents. It aims to make knowledge contained in applications explicit. Thus ontology offers a direction towards solving the semantic interoperability problem and allows computers to operate automatically.

There are several ontology languages available to encode ontology. OWL (Web Ontology Language) is the most popular ontology language. We will introduce the details of OWL in the next chapter. Here we just introduce its basic concepts. In general, OWL is a standard-based ontology language for making ontological statement. OWL was developed based on RDF (Resource Description Framework), RDFS (RDF Schema), OIL (Ontology Inference Layer), DAML (DARPA Agent Markup Language), and DAML+OIL. OWL has been endorsed by the World Wide Web Consortium (W3C) and aims to be used over the World Wide Web. All elements (classes, properties, and individuals) of OWL are defined as RDF resources, and identified by URIs (Uniform Resource Identifiers).

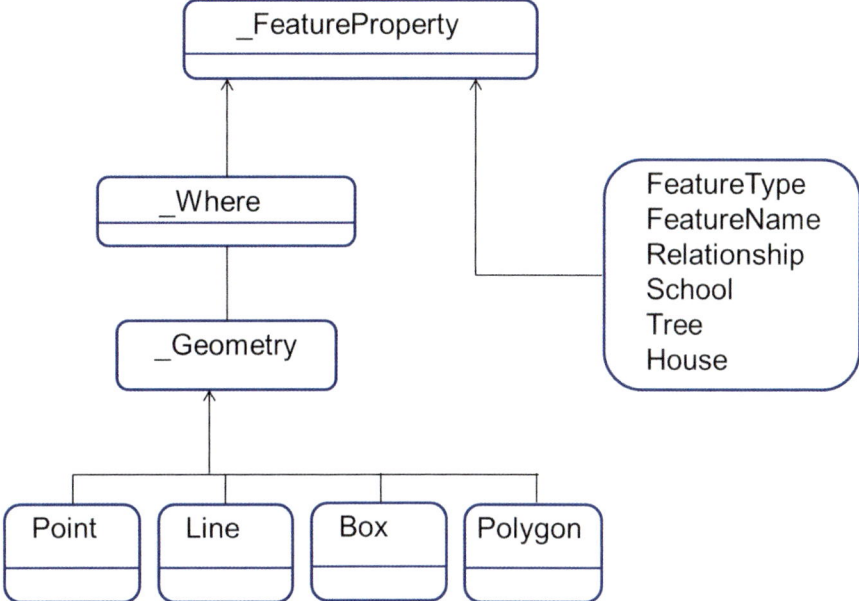

Fig. 2.3 The GeoRSS feature model

Geospatial ontology represents geospatial concepts and properties for use over the Internet. Efforts had been made to develop geospatial ontologies for use on the World Wide Web. For example, the W3C Incubator group on geospatial ontologies had developed geospatial foundation ontologies (http://www.w3.org/2005/Incubator/geo/XGR-geo-ont-20071023/). In general, Geospatial ontology includes geospatial features and geospatial relations. Geospatial features include *Point*, *Line*, and *Area* concepts. Please note that the terms "*Point*", "*Line*", and "*Area*" here refer to types of geospatial features, and they are not geometries of geospatial features. Geospatial relations include *equal*, *disjoint*, *intersect*, *touch*, *cross*, *within*, *contain*, *overlay*, *near*, *connected*, *in front of*, and *around*. We will introduce geospatial relations in details in the next chapter.

Geospatial ontologies adopt the *GeoRSS* feature model, which allows descriptions of *rectangles*, *points*, *lines*, and *polygons* as geometric representation properties of discerned geographic features. This is slightly different and substantially reduced from the OpenGIS *Simple Features* model, in which the *Simple Features'* geometric representations consist of *points*, *multipoints*, *curves*, *multicurves*, *surfaces (polygons)*, and *multisurfaces*. Figure 2.3 shows the *GeoRSS* feature model. From the figure it can be seen that the *GeoRSS* feature model provides a general feature property (**_featureproperty**) to be used to characterize any appropriate content as a geographic feature. Specific subproperties such as **geo:where** can be used to associate the geospatial feature with one of the geometric types: *Point*, *Line*, *Box*, and *Polygon*. **Geo:where** is the general geometric property of the *GeoRSS* GML. A *Point* has a single coordinate pair. A *Line* has two or more coordinate pairs. A *Box* is generally used to roughly delineate an area within which other data lie. A *Box* has exactly two coordinate pairs. A *Polygon* has at least four Coordinate pairs. Other subproperties such as *feature type* and *feature name* can be used to describe other commonly used feature attributes.

The *GeoRSS* feature model is consistent with the ISO standards. However, it provides a subtle difference in emphasizing Web-like feature view or aspect instead of a database-like content. Geo OWL (http://www.w3.org/2005/Incubator/geo/XGR-geo/W3C_XGR_Geo_files/geo_2007.owl) provides an ontology that utilizes the existing *GeoRSS* vocabulary for geographic properties and classes.

2.3.2   Semantic Descriptions of Geographic Information Using Ontology

To realize automatic search and discovery of geospatial feature data at the semantic level, one important challenge is how to match geospatial features to the predefined geospatial ontology. There are two methods to do so: (1) manual match and (2) automatic match. It is time consuming to manually match geospatial features to the predefined geospatial ontology. However, it is not easy to automatically match geospatial features to the predefined geospatial ontology. In the following paragraphs, we describe a supervised machine learning method to realize automatic match after a few initialization steps.

The idea of using the supervised machine learning techniques to automate this process is to develop a matching tool with machine learning components to identify potential matching candidates. Using the matching tool users should be able to make adjustments to correct the matching results. The basic steps of the supervised machine learning method are the followings:

1. Define domain ontology concepts.
2. Create a training data set by labeling the selected geospatial schemas with the correct ontology concepts.
3. Determine the feature representation of the learned function that maps ontology concepts to geospatial schemas. The features may include *geometry types*, and *property names*, *values*, and *data-types of geospatial schemas*. Other features such as *bounding box*, *density*, and *geometry relations* can also be considered.
4. Determine the structure of the learned function and the corresponding learning algorithm. A decision tree learning (Berikov and Litvinenko 2003) can be applied for this purpose since it is able to handle categorical data (e.g. geometry types such as points, lines, and polygons).
5. Apply the learning algorithm to the training data set to obtain parameters, which will be adjusted to maximize performance on the test data sets.

As an example, suppose we have domain ontologies for a transit system and we also have a collection of web features of transit data such as *streets*, *bus routes*, *bus stops*, *patterns*, *time-points*, *trips*, *schedules*, and *facilities*. These web features

Fig. 2.4 This is an example
of using a decision tree for
classifying the web features
into several ontology classes.
The dependent variable is the
number of the web features
in each ontology class, which
can be *Route*, *Stop*, *Street*,
or *Facility*. The independent
variables are *geometry type*
and *density*

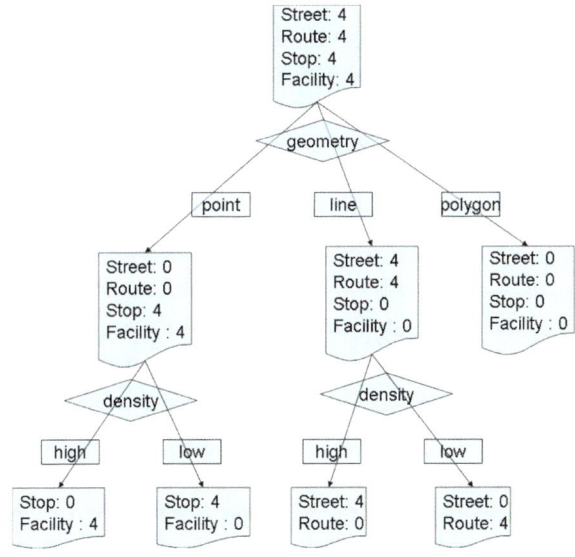

are used by many regions with some differences and similarities. First, we select
a subset of web features and manually map them to the ontological concepts as a
set of training data. Then we run a machine learning algorithm to derive a matching function *f* based on the training data. Finally, we apply the matching function
f to the rest of the web features. To generate the matching function *f*, we need to
decide how to match ontology classes, how to match ontology properties, and how
to generate ontology individuals. For example, to decide how to match web feature
types to ontology classes, we can use the *feature geometry (type and density), feature type name*, *feature property list*, and *property values* as independent variables.
The ontology class that a web feature should match is considered to be a dependent
variable. A decision tree learning algorithm can be used to classify the web features
into each ontology class.

Figure 2.4 shows an example of using a decision tree to classify the web features
into the ontology classes of *Street*, *Route*, *Stop*, and *Facility*. We start with four web
features for each class at the top node of the decision tree. We have two independent variables—*geometry type* and *density*. Through the learning algorithm, we find
the sub-tree of *point* geometry includes all the web features for *Stop* and *Facility*
classes while the sub-tree of *line* geometry includes all the features of *Street* and
Route classes. There is no feature for the sub-tree of *polygon* geometry as expected.
Since there are much more *bus stops* than *facilities* in this example, the predicative
attribute of *geometry density* can be used to separate features of *Stops* from those of
Facilities. Similarly, *Street* class contains features with a higher density than *Route*
class does. We can also use a similar strategy to determine how to match the web
feature properties with the ontology properties. For the web feature properties, we
may consider the *web feature type*, *data-type*, *property value*, and *property name*.

The advantage of the machine learning approach is that it can identify the hidden semantic links between the existing ontology concepts and the geospatial schemas of the legacy data. Though it may not resolve all semantic ambiguities, the supervised machine learning method can potentially extract the maximum amount of information from the training data, which may be annotated with helps of the domain experts. A more precise alternative is to manually map ontological concepts to all geospatial schemas, but this approach is time-consuming, error-prone, inconsistent, and not adaptable to the changing requirements.

2.3.3   The Ontology-based Catalogue Service for Service Discovery

It is important to find suitable geospatial information from the open and distributed environment of the current geospatial web services. Geospatial web service discovery is a process of locating or discovering one or more documents that describe a particular geospatial web service. Although service brokers (or catalogue services) provide searchable repositories of service descriptions via metadata, the traditional service brokers are still insufficient for automatic service discovery based on data contents. The keyword-based search approaches perform syntactic matching, that is, they retrieve service descriptions that contain particular keywords from the users' query. This often leads to poor discovery results because the keywords in the query may be semantically similar but syntactically different, or syntactically similar but semantically different from the terms in a web service description (Broens et al. 2004). For example, "*company*" and "*firm*" (synonyms) are semantically similar but syntactically different, while "*by*" may refer to "*pass a given point*" under some situations or "*not for immediate use*" under other situations (homonyms), thus being syntactically equivalent but semantically different from the terms in the service descriptions. The traditional keyword-based search approaches, therefore, are inherently restricted by the ambiguities of natural language. Ontology provides a possible approach to overcome the semantic heterogeneity problem by identifying and associating semantically corresponding concepts. Ontology-based geospatial web service descriptions reduce ambiguities caused by natural language and are machine-interpretable; thus they make automatic service discovery possible.

Once schemas and attributes of geospatial features are mapped to the ontologies, the contents of geospatial web services can be searched and utilized effectively over the Web. An ontology server can be used by the service broker to map the standard OGC catalog services to the ontology-based catalog services. Through a common ontology server to define a common semantic meaning of web services, the ontology-based catalog can explicitly specify traditional web services at semantic level and support automatic discovery, composition, invocation, and orchestration of the data services from the diverse sources. Unlike the traditional web service broker such as the OGC catalog service that enables discovery and retrieval of metadata, the ontology-based catalog service in the framework provides searchable repositories of service descriptions at semantic level and allows users to search and access geospatial information directly based on semantic contents. Instead of searching the resource-based metadata that describes web services and data sets, users can directly locate, access, and make use of contents of geospatial services and data sets in a distributed system through the ontology-based catalog service.

The ontology-based catalog service is able to capture semantics of the users' query, the web services, and the contextual information that is considered relevant in the matching and discovering process. The ontology-based catalog service has advantages for the following three reasons (Broens et al. 2004): (1) It provides a *vocabulary* for modeling knowledge in a restricted domain. Because the vocabulary is built via a consensus within a community of interest, it enables the seamless knowledge interchange. (2) It is usually grounded with formal semantics such as model theory or description logic; thus it enables unambiguous definitions of compound concepts. Based on these definitions it is possible to infer new implicit information from explicit information. (3) Since the information provided by ontologies can be understood not only by humans but also by computers, it is possible to perform automatic information processing through the ontologies.

The following section introduces a set of geospatial feature discovery algorithms for automatic discovery of geospatial features based on the ontology-based WFS, recently proposed by us (Zhang et al. 2010a).

Geospatial Feature Discovery Algorithms

To find feature-level geospatial data, we consider a WFS feature as a basic unit of service rather than the entire WFS server. The reason is that a user should not be concerned with whether two features are located in different servers. Instead, a query processing module should handle the task of joint query transparently. In this way, requesters can search separate sources that provide the requested information. The discovery algorithms consider a WFS description as a consistent collection of restrictions over the named properties of a WFS, such as *URI of the WFS server*, *Feature Type name*, *Feature Property name*, *Geometry Type of Feature*, and *Bounding Box of Geometry*.

Definition 1 All WFS features are represented as tuples with the form (T, P, G, B), where T is the Feature Type name, P is the Feature Property name, G is the Geometry Type of the Feature, and B is the Bounding Box of the Geometry. Let $Q = (T', P', G', B')$ be a service query. The discovery problem can be defined as automatically finding a set S of WFSs such that $S = \{(T, P, G, B) \mid T <: T', P <: P', G <: G', B <: B'\}$, where $<:$ defines a partial order on features, property sets, geometry types, and bounding boxes.

Define $T_1 <: T_2$ **if** T_1 **corresponds to an ontology class that is the same as or a subclass of** T_2.

Define $p_1 <: p_2$ **if feature property** p_1 **corresponds to an ontology property that is the same as or a subproperty of** p_2.

Fig. 2.5 Geospatial feature discovery
process. (Source: Zhang et al. 2010a)

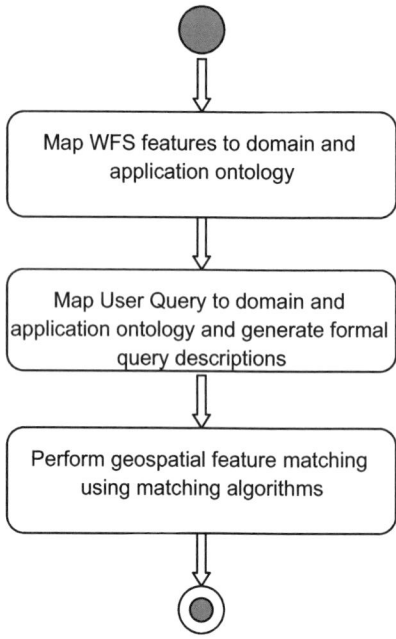

Define $P_1 <: P_2$ if for each feature property p_2 in P_2, there exists a p_1 in P_1 such that $p_1 <: p_2$. We may relax this definition to replace "for each" with "for some" to allow more matches.

Define $G_1 <: G_2$ if G_1 is a geometry type that is equal to or a subtype of G_2.

Define $B_1 <: B_2$ if B_1 is a bounding box contained in B_2. We may relax this definition so that the partial order holds even if they only intersect to retrieve more feature results.

Before the discovery process, we need to index the available WFS features in an index file for a more efficient search when there is a large number of features. The indexing process is to check all available services and collect the following fields for each feature: (1) *URI of the WFS server*; (2) *Feature names*; (3) *Feature Property names*; (4) *Geometry Type of Features*; (5) *Bounding box of Geometry*. A service broker, responsible for collecting WFSs, maintains the index file with an entry for each WFS feature.

The discovery process has several steps as illustrated in Fig. 2.5. First, we map each WFS feature to its domain and application ontologies using the algorithm illustrated in Fig. 2.6a. We use the breadth-first search algorithm to traverse the ontology class hierarchy to search for the matched ontology classes. Leaf classes are searched first, then their immediate superclasses, and so on. For instance, if one WFS bus route feature has the term *Route*, we would map it to *TransitRoute* ontology instead

(A) Algorithm for Mapping WFS features to domain and application ontology

1   **Input**   available WFS features F
2   domain and application ontology O
3   **Initialize** I = empty set

4   **for each** feature f in F
5   find a class T in O that matches the feature name
6   find a property set P in O such that each of P matches a feature property
7   find a class G in O that matches the geometry type of f
8   create an instance B of bounding-box class in O that corresponds to the bounding box of f
9   add (T, P, G, B) to I
10 **end for**

11 **Output** an index file I

(B) Algorithm for generating formal query descriptions

1   **Input**   target query q
2   domain and application ontology O

3   if q specifies a feature name, then find a class T' in O that matches the name, else let T' be the root Feature class in O
4   for each property specified in q, find a matching property in O and collect the properties into a set P'. Otherwise, P' is empty.
5   if q specifies a geometry name, then find a matching class G' in O, else let G' be the root Geometry class in O
6   if q specifies a bounding box, then create an instance B' of the bounding-box class in O, else let B' be the largest bounding box.

7   **Output** query in form of (T', P', G', B').

(C) Algorithm for geospatial feature match

1   **Input**   index file I of WFS
2   feature query (T', P', G', B')

3   **Initialize** I' = empty set

4   **for each** (T, P, G, B) $\in$ I
5   if T <: T' and P <: P' and G <: G' and B <: B', then add (T, P, G, B) to I'
6   **end for**

7   **for each** $i = (T,P,G,B) \in I'$
8   compute $s(i) = S_T(T,T') * W_T + S_P(P,P') * W_P + S_G(G,G') * W_G + S_B(B,B') * W_B$
9   **end for**

10 Rank each element i in I' using s(i) so that i is before j if $s(i) > s(j)$.

11 **Output** a set of WFS features I'

Fig. 2.6 Geospatial feature discovery algorithms. (Source: Zhang et al. 2010a)

of its superclass, *Line*. Because the algorithm uses breadth-first search, it is guaranteed that the more precisely matched subclass in the hierarchy has already been checked when the more general superclass is being compared.

Second, we map a user's query to domain and application ontologies and generate formal query descriptions using the algorithm shown in Fig. 2.6b. The discovery engine abstracts a user's query into four parts of service descriptions that are required to answer the query: *Feature Type names, Feature Property names, Geometry Type of Features*, and *Bounding Box of Geometry*. These four parameters of capability descriptions must reflect (1) the semantic contents of the query and (2) the requirements of the generated requests.

Third, find the matched WFS features using the matching algorithm in Fig. 2.6c. The discovery engine finds the appropriate WFS features by matching the descriptions required to solve the query with the descriptions of providers through the parameters. The algorithm first uses the query *Bounding Box of Geometry* parameter to narrow down the list of services in the repository. It acquires all of those services that produce at least matched *Bounding Box of Geometry* (all WFSs that are located within the geography limitation). From those services, it further narrows down the list of services by *Geometry Type of Features*, then by *Feature Type names*, and finally by *Property names*. All the description parameters provided by WFSs must be equivalent to or subsume the required description parameters in the query. Whenever an exactly equivalent match is found, it is recorded with the highest score. Otherwise, according to the degree of the match detected, it is recorded with a lesser score. We calculate the score using the following formula:

$$s(i) = S_T(T,T') * W_T + S_P(P,P') * W_P + S_G(G,G') * W_G + S_B(B,B') * W_B \quad (2.1)$$

where $s(i)$ is the calculated score, S_T is the similarity score between two *ontology feature classes* T and T', S_P is the similarity score between two *feature property* sets, S_G is the similarity score between two *Geometry Type of Feature* sets, S_B is the similarity score between two *Bounding Box of Geometry* sets, and W_P, W_G, W_B are the weights of *the property set, Geometry type* and *Bounding box*, respectively. The similarity score S_T of two ontology feature classes, T and T', can be based on the number of inheritance levels between them. For example, $S_T(T,T') = 0$ and $S_T(T,T') = 1$ if T is a direct subclass of T'. W_T is the weight of the similarity score of ontology feature classes. The similarity score S_P of two property sets is computed the same way except that we add up the similarity scores between the matched properties in the two sets to produce the similarity score of the two sets. S_G is computed the same way as S_T. S_B is calculated as below:

$$S_R(B,B') = size\,Of\,(overlap\,Area\,Of\,(B,B')) \quad (2.2)$$

The main advantage of the matching algorithm is that it supports a flexible semantic match between the provided WFSs and the requests. The match between the descriptions of the requester and the descriptions of the WFS provider depends on the relation between the concepts associated with those parameters in descriptions.

For instance, consider how a request with the feature class *MilwaukeeBusRoute* matches a WFS when the feature class is *TransitRoute*. Given a transit ontology, the discovery engine would match *MilwaukeeBusRoute* with *TransitRoute* instead of *Highway*, because *MilwaukeeBusRoute* is a subclass of *TransitRoute*, while it has no direct relation with *Highway*.

Furthermore, the result of the match is not a hard "true" or "false", and it depends on the degree of similarity between the concepts in the match. The discovery engine can draw an inference between descriptions of the provided WFSs and requests on the basis of available ontologies. Despite the flexibility, the discovery engine still rejects WFS features that do not match the requests, and accepts, but with a low score, matches that may not be satisfactory for the requester.

To evaluate the accuracy of the algorithm, we set up a benchmark set of web features and a suite of test cases. Each test case is a query that can be answered by one or more web features in the benchmark set. We first identify the web feature types that contain answers to our test cases, and then use them as ground truth to our test cases. The evaluation of a query uses two metrics—*precision* and *recall*, as follows:

$$Precision = \text{\# of correct feature types returned by the algorithm / total}$$
$$\text{\# of feature types returned by the algorithm}$$
$$Recall = \text{\# of correct feature types returned by the algorithm / total}$$
$$\text{\# of correct feature types in the benchmark set}$$

Precision is related to the false positive rate while recall is related to false negative. Higher precision can be achieved through a stricter matching standard in the algorithm, but it can decrease recall by potentially missing more correct answers. Part of the evaluation process is to find suitable parameters for the algorithm to achieve reasonable balance of precision and recall.

2.3.4   Web Service Composition

To utilize web services effectively, users have to analyze web services and evaluate their applicability to the task at hand dynamically. Under some situations, one web service may be not enough and users may need to combine several web services dynamically in a certain sequence at run-time to complete a task need. The dynamic nature of the availability of web services and the large number of alternative combinations of service choices make the dynamic service composition a formidable task. A key challenge in promoting widespread use of web services in the geospatial applications is to automate construction of a chain or process flow that involves multiple services and highly diversified and distributed data (Yue et al. 2007). Ontology-based geospatial semantics may be used for enabling the automatic discovery, access, and chaining of geospatial web services. According to Yue et al. (2007), the key to achieve automation of web service discovery and composition relies mainly on solutions to three issues: (a) make geospatial web services interoperable

Fig. 2.7 Web services composition process. (Source: Zhang et al. 2010b)

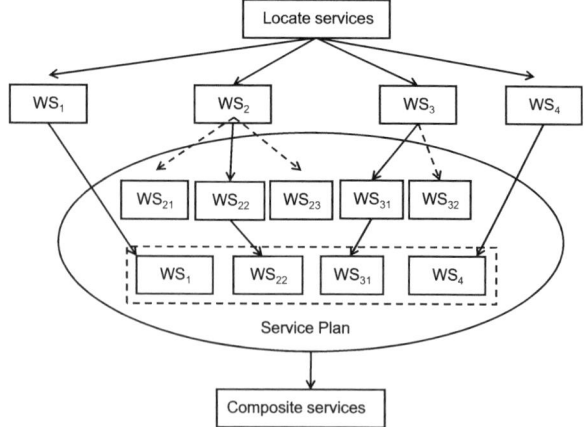

both syntactically and semantically; (b) based on the syntactic and semantic descriptions, automatically discover the most appropriate data and services; and (c) assemble the most appropriate data and services to build the composite service (Di 2005). The ontology-based SOA (Service-Oriented Architecture) is the basis for automatic service composition. The service management functions such as registration, discovery, accessing and execution in SOA are needed for the automation process of web service composition. While syntactically interoperable web services can be composed manually in the SOA, semantically interoperable web services can be composed automatically in the ontology-based SOA.

Two steps may be needed in facilitating dynamic service composition over geospatial semantic web: (1) Locate possible services based on adequate descriptions of the users' request. The rich semantics added to web service descriptions by OWL-S enable semantic composition by matching service capability descriptions at semantic level to requirements. (2) Based on the necessary control and data flow constraints among services, develop a service plan to invoke the execution among the services in the correct order. The service plan can be created and updated dynamically. The service plan can be created based on the centralized broker that manages the service composition process. The service plan can determine which service requests can be feasibly processed and which requests cannot be fulfilled. When one web service cannot satisfy a user's request, the service plan will communicate and cooperate with other services to finish the composition service for the request. Thus, the geospatial semantic web should exhibit the systemic characteristics such as self-organization, evolution, scalability, and adaptability, although each of the geospatial web services is not designed to do so.

Figure 2.7 shows the web services composition process. It is based on two processes—service discovery process and service composition process. Service discovery is based on semantic service descriptions. Service discovery automatically determines web services that will precisely fulfill the user's needs. If a single data service cannot be found, the service search engine will search for two or more services that can be composed to synthesize the required service. This is called web service composition.

The following section introduces a geospatial web feature service composition algorithm we recently proposed for automatic integration of WFS (Zhang et al. 2010a).

Geospatial Web Feature Service Composition Algorithm

Should the matching engine not find a single WFS that matches the user's query, it will search for two or more WFSs that can be composed to synthesize the required service using composition algorithms. This task is called composition.

Definition 2 Let P be the set of all provided WFS features in a given web service repository. A WFS is represented as tuples with the form (T, P, G, B), where T is a Feature Type name, P is a set of Feature Property, G is the Geometry Type of a Feature, and B is the Bounding Box of a Geometry. Let $Q=(T', P', G', B')$ be a service query. The composition problem can be defined as automatically finding a set S of Services such as $S=(S1, S2, \ldots Sn)$ where for all i, $Si=(Ti, Pi, Gi, Bi)$ and

$$T_i <: T', (P_1 \cup P_2 \cup \ldots \cup P_n) <: P', G_i <: G', (B_1 \cup B_2 \cup \ldots \cup B_n) <: B'$$

When no completely matched WFS is discovered in order to satisfy a user's objective, the existing partially matched WFSs may need to be combined together to fulfill the query request. The WFS composition is the process of selecting and combining WFSs to achieve the user's goal, which cannot be realized by the existing WFSs. To perform automated composition, the steps in the discovery process will be performed first in order to allow the reasoning system in the discovery engine to order and combine WFSs. If the discovery engine cannot find an existing appropriate atomic WFS to satisfy a user's objective, it will try to find partially matched WFSs or composable services and combine the partially matched WFSs or composable services to achieve the user's objective by using the composition process and algorithm as illustrated in Figs. 2.8 and 2.9. As shown in Fig. 2.8, in order to produce the composition service, the composable WFSs that are useful for the composition are selected at multiple stages: First, the discovery engine tries to retrieve the WFSs with *Feature Type name T* and *Geometry Type of the Feature G* parameters such that T ontology is equal to or a subclass of the query ontology T' and G ontology is equal to or a subclass of the query ontology G'. This will result in a set of service S_1. Second, from the set of service S_1, the discovery engine narrows down the subset S_1 to a subset S_2 under the condition that the feature property P is equal to or a sub-property of query feature property P', where P is the union of the property parameters of the WFSs in S_2. Third, the discovery engine further narrows down the subset S_2 to a subset S_3 under the condition that the bounding box B is contained in the query bounding box B', where B is the union of the bounding box parameters of the WFSs in S_3. We repeat step two and three until S_3 is found or all possible S_2's

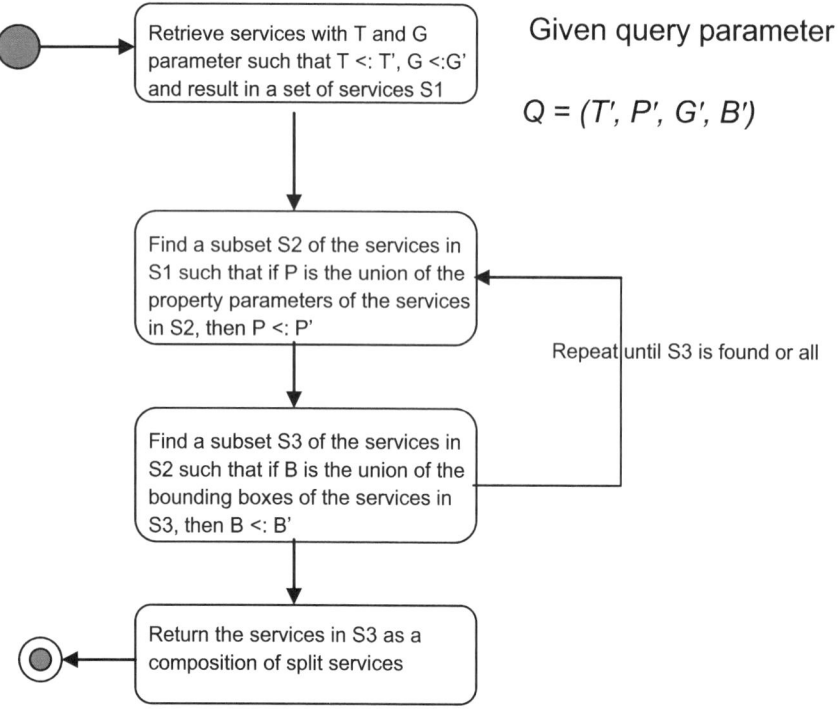

Fig. 2.8 WFS composition process. (Source: Zhang et al. 2010a)

have been tested. Finally, we return the services in S_3 as a composition of the split services. The formal algorithm for service composition is shown in Fig. 2.9.

The evaluation of the query composition algorithm uses the same benchmark set and test suite used for evaluating the feature discovery algorithm. However, the ground truth for the test suite is some sets of web feature types that need to be combined to give complete answers to the original query. We also use *precision* and *recall* to evaluate the composition algorithm. Again, there could be tradeoff between improving *precision* and improving *recall*. We can increase *recall* by imposing more relaxed rules in selecting web feature types, but this could potentially decrease *precision*.

2.4 Chapter Summary

This chapter introduces background information about the semantic problem of spatial data. Although the local, regional, and global SDIs have undoubtedly improved sharing and synchronization of geospatial information across the diverse resources, there are limitations of the currently implemented SDIs and it might still be difficult to find data sources from the currently implemented SDIs. A difference in the

1	**Input**: Q = (T′, P′, G′, B′)
2	A set of WFS S
3	**Initialize**: let sets S1, S2, and S3 be empty
4	**For** each s = (T, P, G, B) in S
5	If T <: T′ and G <: G′, then add s to S1
6	**End for**
7	A property p′ is *covered* by a service s = (T, P, G, B) if there is a p in P such that p <: p′.
8	A bounding box B′ is *covered* by a set of services if the union of the bounding boxes of the services covers B′
9	**Repeat until** either B′ is covered by services in S3 or all possible S2 has been tested
10	**Set** S2 to empty and let S1′ = S1
11	**Repeat until** either S1′ is empty or all properties in P′ are covered by services in S2
12	Take a service s = (T, P, G, B) from S1′ such that P covers at least one property in P′ that are not already covered by services in S2
13	**Add** s to S2
14	**Set** S3 to empty
15	**Repeat until** either S2 is empty or B′ is covered by services in S3
16	Take a service s = (T, P, G, B) from S2 such that B and B′ overlaps
17	**Add** s to S3
18	**Output**: The services in S3 as a composition of split services (that can be executed in concurrently)

Fig. 2.9 WFS composition algorithm. (Source: Zhang et al. 2010a)

semantics used in diverse data sources is one of the major problems in spatial data sharing and data interoperability. Geospatial Semantic Web was recently proposed in geospatial community to overcome the semantic heterogeneity problem of geospatial data. Geospatial Semantic Web provides computers meaningful geospatial contents, thus it allows geospatial data to be discovered, queried, and consumed automatically by computers. The systems built on the Geospatial Semantic Web technologies can automatically search and access geospatial data by their contents rather than just by keywords in metadata. In this chapter we introduced the main technologies used in a Geospatial Semantic Web architecture—(1) *ontology*, (2) *semantic descriptions of geographic information using ontology*, (3) *the ontology-based catalogue service*, (4) *web service discovery*, and (5) *web service composition*.

References

Askew D, Evans S, Matthews R et al (2005) MAGIC: a geoportal for the English countryside. Comp Environ Urban Syst 29:71–85
Berikov V, Litvinenko A (2003) Methods for statistical data analysis with decision trees. Sobolev Institute of Mathematics, Novosibirsk

Berners-Lee J, Hendler J, Lassila O (2001) The semantic web. Sci Am 184:34–43

Bishr Y (1998) Overcoming the semantic and other barriers to GIS interoperability. Int J Geogr Inf Sci 12:299–314

Broens T et al (2004) Context-aware, ontology-based service discovery. In: Markopoulos P et al (eds) EUSAI 2004, LNCS 3295. Springer-Verlag, Berlin, pp 72–83

Crompvoets J, Bregt A, Rajabifard A et al (2004) Assessing the worldwide developments of national spatial data, Inter J Geogr Inf Sci 18:665–689

Di L (2005) A framework for developing web-service-based intelligent geospatial knowledge systems. Geogr Inf Sci 11:24–28

Egenhofer MJ (2002) Toward the semantic geospatial web. In: Proceedings of the Tenth ACM International Symposium on Advances in Geographic Information Systems, McLean, Virginia, 8–9 Nov 2002

Farrugia J, Egenhofer MJ (2002) Presentations and bearers of semantics on the web. In: Proceedings of the Fifteenth International Florida Artificial Intelligence Research Society Conference (FLAIRS 2002), Pensacola, Florida, pp 408–412

Gruber TR (1993) A translation approach to portable ontology specifications (PDF). Knowl Acquis 5:199–220. doi:10.1006/knac.1993.1008

Lutz M (2007) Ontology-based descriptions for semantic discovery and composition of geoprocessing services. Geoinformatica 11:1–36

Lutz M, Kolas D (2007) Rule-based discovery in spatial data infrastructure. Trans GIS 11:317–336

Mansourian A, Rajabifard A, Valadan Zoej MJ et al (2006) Using SDI and web-based system to facilitate disaster management. Comp Geosci 32:303–315

Masser I (2005) GIS worlds: spatial data infrastructures. ESRI Press, Redlands

Peng ZR, Zhang C (2004a) The roles of geography markup language, scalable vector graphics, and web feature service specifications in the development of internet geographic information systems. J Geogr Syst 6:95–116

Peng ZR, Zhang C (2004b) GML, WFS, SVG, and the future of internet GIS. GIS@ Dev 8(7):29–32

Peng ZR, Zhang C (2005) A new trend of internet GIS development: geospatial semantic web based on services-oriented architecture. GIS@ Dev 9(10):34–37

Rajabifard A, Binns A, Williamson I (2006) Virtual Australia: developing an enabling platform to improve opportunities in the spatial information industry. J Spat Sci 51:63–78

Tait MG (2005) Implementing geoportals: applications of distributed GIS. Comp Environ Urban Syst 29:33–47

Wiegand N, García C (2007) A task-based ontology approach to automate geospatial data retrieval. Trans GIS 11:355–376

Williamson I, Rajabifard A, Feeny M-E (2003) Developing spatial data infrastructure, from concept to reality. Taylor and Francis, New York

Yue P, Di L, Yang W et al (2007) Semantics-based automatic composition of geospatial web service chains. Comp Geosci 33:649–665

Zhang C, Li W (2005) The roles of web feature service and web map service in real time geospatial data sharing for time-critical applications. Carto Geogr Info Sci 32:269–283

Zhang C, Li W, Peng ZR et al (2003) GML-based interoperable geographical databases. Cartography 32:1–16

Zhang C, Li W, Zhao T (2007) Geospatial data sharing based on geospatial semantic web technologies. J Spat Sci 52:11–25

Zhang C, Zhao T, Li W et al (2010a) Towards logic-based geospatial feature discovery and integration using web feature service and geospatial semantic web. Inter J Geogr Info Sci 24:903–923

Zhang C, Zhao T, Li W (2010b) A framework for geospatial semantic web based spatial decision support system. Inter J Digital Earth 3:111–134

Chapter 3
Ontology languages and Geospatial Semantic Web

3.1   RDF (Resource Description Framework)

RDF (Resource Description Framework) is a family of W3C specifications origi-nally designed as a metadata data model. It has been used as a general method for conceptual description or modeling of information in the Web through a Graph-based data model. RDF isn't strictly an XML format, and it's also not just about metadata. It is a general method to decompose any type of knowledge into small pieces, with some rules about the semantics or meaning of those pieces. The core structure of the RDF is a set of triples, which is known as RDF graphs and consists of three components—a subject, a predicate, and an object. Figure 3.1 illustrates the three components in an RDF graph, which can be represented as a node-arc-node diagram. "Subject" and "Object" can be represented as "nodes" while "Predicate" can be represented as "arcs".

RDF allows making statements about resources with the following structure:

<subject> <predicate> <object>

Where the "subject" and the "object" represent the two resources being related and the "predicate" represents the nature of the relationship between the two re-sources. The relationship is also called a "property" in RDF. Because RDF state-ments consist of three elements, they are also called RDF "triples". The followings are some examples of RDF triples:

<UConn> <is_a> < university>

<UConn> <east_of> <Hartford >

<Professor Chuanrong Zhang> <professor_at> <UConn>

<Professor Chuanrong Zhang> <teaches> <ArcGIS software>

<ArcGIS software> <developed_by> <the ESRI company>

C. Zhang et al., *Geospatial Semantic Web,* DOI 10.1007/978-3-319-17801-1_3

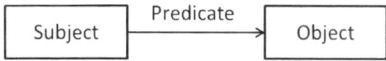

Fig. 3.1 An RDF graph with two nodes (*subject and object*) and an arc connecting them (*predicate*)

As shown in the above example, the same resource can be referenced in multiple triples. The important part of RDF's power is that it is possible to find connections between triples of RDF because RDF has capability to have the same resource be in the subject position of one triple and the object position of another triple.

The RDF triples can be visualized by a connected RDF graph, which consists of nodes and arcs. A graph is basically a network. RDF as a graph expresses exactly the same information as RDF written out as triples, but the graph form makes it easier for users to see structure in the data. RDF allows users to group RDF statements in multiple graphs and to associate such graphs with an IRI (International Resource Identifier). Figure 3.2 illustrates the RDF graph resulting from the aforementioned sample triples. Once a connected RDF graph has been created, SPARQL (SPARQL Protocol and RDF Query Language), which is an RDF query language, can be used to query the needed information from the RDF databases. SPARQL is introduced in detail in the next chapter. RDF allows users to combine triples from any source into a graph and process it as legal RDF. Thus there are many datasets on the Web using interlinked RDF, and many of them offer a query capability through SPARQL. For example, Wikidata and WordNet are two examples of such datasets.

There are three types of nodes in an RDF graph—IRIs, literals, and blank nodes. IRIs and literals are also called "resources" or "entities" as used in the RDF semantic specification, and they can represent something in the world. The resource denoted by an IRI is known as its referent, and the resource denoted by a literal is known as its literal value. The relationship between resources, as indicated by the "predicate", is also an IRI known as a "property".

Fig. 3.2 A connected RDF graph of the sample triples

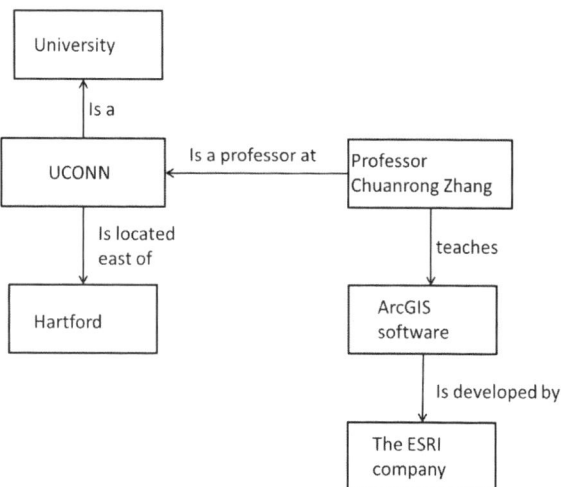

The IRI is abbreviated from "International Resource Identifier". The URLs (Uniform Resource Locators) that have been used as Web addresses are one form of IRI. Other forms of IRI offer an identifier for a resource without acknowledging its location or how to access it. IRIs can appear in all three positions (a subject, a predicate, and an object) of a triple. IRIs can be used to find many different types of resources such as documents, people, physical objects, and abstract concepts (Schreiber and Raimond 2014). Because IRIs are global identifiers, users can reuse them to identify the same thing.

Literals are basic values that are not IRIs (Schreiber and Raimond 2014). Literal values are raw text that can be used instead of objects in RDF triples. Unlike names (i.g. URIs) which represent things in the real world, literal values are just raw text data inserted into the graphs. For example, number "1.2356" may be a literal. For another example, a date such as "the 21th of March, 2014" may be also a literal. Unlike IRIs, literals may only appear in the "object" position of a triple. Literal values can be optionally adorned with two pieces of metadata: "language tag", to specify which language the raw text is written in, and "datatypes", to indicate how to interpret the raw text. Many "datatypes" defined by XML Schema are literals such as string, Boolean, integer, decimal and date. Literals associated with a "datatype" enable such values to be parsed and interpreted correctly (Schreiber and Raimond 2014).

A blank node is a resource without a global identifier, and it is also called an anonymous resource. The word 'anonymous' or 'blank' means that these are nodes in a graph without a name. Unlike IRIs and literals, blank nodes do not represent specific "resources"; they just indicate something with a given relationship. Blank nodes are similar to simple variables in algebra, which represent something without specified values. A blank node can be the "subject" or the "object" of a RDF triple. It represents a resource without explicitly naming it with an IRI.

Although the RDF data model provides a way to make statements about resources, it makes no assumptions about the resources that IRIs stand for. Therefore, RDF usually has to combine with vocabularies or other conventions to provide semantic information about these resources. The RDF Schema language (Brickley and Guha 2014) provides the definitions of vocabularies for RDF. The RDF Schema language allows users to define semantic characteristics of RDF data.

In RDF Schemas, "class" is used to specify categories of resources. A class is a type of thing. For example, "Professor Chuanrong Zhang" is a member of the class "Person". For another example, "ArcGIS software" is a member of the class "Software". Classes are themselves members of the type "Class". An interesting class of RDF is "rdf:property". Any entity used as a predicate is an "rdf:property". The "rdf:type" property is used to identify the relation between an instance and its class in a RDF Schema. This predicate is used to indicate what kind of things a resource is. The "rdf:type" can relate an entity to another entity that denotes the class of the entity. "rdf:domain" and "rdf:range" restrictions can be used to define type restrictions on the subjects and objects of particular triples. With RDF Schemas, users can create hierarchies of classes and sub-classes and those of properties and sub-properties.

RDF is designed to represent knowledge in a distributed world. By conveying information in the same simple style, that is, the subject-predicate-object triple, RDF is able to automatically merge useful information from multiple sources to form a larger coherent and useful collection. Using the simple model, RDF allows structured and semi-structured data to be mixed, exposed, and shared across different applications. RDF can facilitate data merging even if the underlying schemas differ.

RDF is particularly concerned with meaning. Everything mentioned in RDF means something—something concrete in the world or an abstract concept, or a fact. With the declarative semantics, RDF is able to make logical inferences. For example, from the status of a certain set of input triples (True or False), RDF can deduce the status of other triples (True or False). Please note that providers and users of RDF data must agree on the semantics of resource identifiers. Although there are some controlled vocabularies in common use, such an agreement is not inherent to RDF itself.

In general, RDF is the W3C standard for encoding knowledge in a distributed world and it provides a standards-compliant method for exchanging and sharing data (Hayes and Patel-Schneider 2014). It is a simple method that can express any fact yet so structured that computer applications can do useful things with it. RDF is the foundation of the semantic web. It adds machine-readable information to Web pages and enriches a dataset by linking it to third-party datasets. Furthermore, RDF can put together RDF files posted by different applications around the Web by linking documents together via the common vocabularies they use and by allowing any document to use any vocabulary. Thus, users can easily learn from them new things that no single document can assert. This is a unique flexibility that RDF has. With RDF, users can integrate data from different sources and reuse data provided by other applications. In addition, with RDF data can be decentralized in such a way that no single party "owns" all the data.

It should be noticed that RDF is designed for knowledge, rather than for data. RDF is particularly concerned with meaning, which can be further defined by using OWL (Web Ontology Language). OWL is built on RDF and provides a language for defining structures, web-based ontologies that enable interoperability of data among different communities. OWL defines more classes, and thus allows RDF authors define more meanings of their predicates within RDF. OWL is introduced in details in the following section.

3.2 OWL (Web Ontology Language)

The OWL (Web Ontology Language) is an ontology language for the Semantic Web with formally defined meaning (Hitzler et al. 2012). OWL ontologies themselves are primarily exchanged as RDF data and can be used along with RDF. But OWL facilitates greater machine interpretability of data content than RDF does through providing additional vocabularies along with a formal semantics (Bechhofer 2009). RDF provides a simple semantics for a data model of "resources" and relations between "resources". RDF Schema provides a vocabulary for describing properties and classes of RDF resources, with a semantics for generalization-hierarchies of such properties and classes. OWL adds more vocabulary for describing properties and classes. OWL provides much more expressiveness than RDF and RDF Schema do. For example, although RDF and RDF Schema provide vocabularies in typed hierarchies (subclass and subproperty relationships, domain and range restrictions, and instances of classes), they provide no vocabularies for describing local scope of properties, disjointness of classes, boolean combinations of classes, cardinality restrictions, or special characteristics of properties (Antoniou and Van Harmelen 2004). OWL adds language primitives to support the richer expressiveness for these features (De Vergara et al. 2004). For example, OWL classes can be specified as logical combinations (intersections, complements, or unions) of other classes, going beyond the capabilities of RDF's. For another example, OWL provides restrictions on how properties behave (that are local to a class). OWL can define classes where a particular property is restricted so that all the values for the property in instances of the class must belong to a certain *class* (or *datatype*). Built on RDF and RDF Schema, OWL allows users to write explicit, formal conceptualizations of domain models.

OWL has three sublanguages: OWL Lite, OWL DL, and OWL Full (McGuinness and Van Harmelen 2004). OWL Lite provides language constructs for a classification hierarchy and simple constraints. OWL DL supports the maximum expressiveness while retaining computational completeness and decidability. OWL Full provides the maximum expressiveness and more syntactic freedom for RDF with no computational guarantees. OWL Lite supports only some of the OWL language features, and it has more limitations on the use of the features than OWL DL or OWL Full has. OWL Full can be considered as an extension of RDF, and OWL Lite and OWL DL can be viewed as extensions of a restricted view of RDF. All of OWL documents (Lite, DL, Full) are RDF documents. All of RDF documents are OWL Full documents but only some RDF documents are legal OWL Lite or OWL DL documents.

OWL is designed to represent rich and complex knowledge about things, groups of things, and relations between things. OWL documents are also known as ontologies, which can be published on the Web (Van Harmelen and McGuinness 2004 2004). "Axioms" are the basic statements that an OWL ontology expresses. "Entities" are elements used to refer to real-world objects in OWL. "Expressions" are combinations of entities to form complex descriptions from basic ones in OWL.

OWL can be used to explicitly represent the meanings of terms in vocabularies and the relationships between those terms. OWL ontologies provide classes, properties, individuals, and data values, and are stored as Semantic Web documents (Motik et al. 2009; Hitzler et al. 2012). In OWL, objects are denoted as "*individuals*", categories are denoted as "*class*", and relations are denoted as "*properties*". "*Properties*" in OWL are further subdivided into three categories—"*Object Properties*",

Table 3.1 The OWL lite language constructs

Categories of OWL constructs	OWL constructs
RDF schema features	*owl:Class; rdfs:subClassOf; rdf:Property; rdf:subPropertyOf; rdfs:domain; rdfs:range; owl:Individual*
(In)equality	*equivalentClass; equivalentProperty; sameAs; different-From; AllDifferent; distinctMembers*
Property characteristics	*ObjectProperty; DatatypeProperty; inverseOf; Transi-tiveProperty; SymmetricProperty; FunctionalProperty; InverseFunctionalProperty*
Property restrictions	*Restriction; onProperty; allValuesFrom; someValuesFrom*
Restricted cardinality	*minCardinality; maxCardinality; cardinality*
Header information	*Ontology; imports*
Class intersection	*intersectionOf*
Versioning	*versionInfo; priorVersion; backwardCompatibleWith; incompatibleWith; DeprecatedClass; DeprecatedProperty*
Annotation properties	*rdfs:label; rdfs:comment; rdfs:seeAlso; rdfs:isDefinedBy; AnnotationProperty; OntologyProperty*
Datatypes	*Xsd datatypes*

Table 3.2 The additional language constructs for OWL DL and OWL full languages

Additional categories of OWL constructs	Additional OWL constructs
Class axioms	*oneOf; dataRange; disjointWith; equivalent-Class; rdfs:subClassOf*
Boolean combinations of class expressions	*unionOf; complementOf; intersectionOf*
Arbitrary cardinality	*minCardinality; maxCardinality; cardinality*
Filler information	*hasValue*

"*Datatype properties*", and "*Annotation properties*". "*Object properties*" are used to relate objects to objects. "*Datatype properties*" are used to assign data values to objects. "*Annotation properties*" are used to encode information about the ontology itself instead of the domain of interest. OWL can express which objects (also called "*individuals*") belong to which classes, and what the property values are for specific individuals.

Table 3.1 lists the OWL Lite language constructs. Table 3.2 lists the additional language constructs for OWL DL and OWL Full languages. The basic OWL language contains *RDF Schema Features, (In)Equality, Property Characteristics, Class Intersection, and Restricted Cardinality. RDF Schema Features* include *Class, rdfs:subClassOf, rdf:Property, rdfs:subPropertyOf, rdfs:domain, rdfs:range, and Individual. Class* defines a group of individuals which share some common properties. *Properties* are used to state relationships between individuals or from individuals to data values. A *domain* of a property limits the individuals to which the property can be applied, while the *range* of a property limits the individuals that the property may have as its values. To attach a property to a class, the tag *rdfs:domain* has to be included. *Individuals* are instances of classes, and *properties* may be used to relate one individual to another. The *equality* or *inequality* feature consists of *equivalentClass, equivalentProperty, sameAs, differentFrom,* and *AllDifferent*, which are employed to state the relationships of classes, properties and individuals. The *equivalentClass* and *equivalentProperty* are used to state the equivalent classes or properties and can be used to create synonymous classes or properties. The *sameAs* and *differentFrom* are used to state the same or different individuals. *Property Characteristics* such as *inverseOf, TransitiveProperty, SymmetricProperty, FunctionalProperty,* and *InverseFunctionalProperty* are used to provide information concerning properties and their values (inverse, transitive, symmetric, and unique). Restrictions *onProperty, allValuesFrom,* and *someVaulesFrom* can be placed on how properties can be used by instances of a class. *Cardinality* restrictions such as *minCardinality, maxCardinality,* and *cardinality* are stated on properties with respect to a particular class. The *intersectionOf* allows intersection of named classes and restrictions. The detail information about OWL syntax is given by W3C website (McGuinness and Van Harmelen 2004).

Please notes that equivalence statements can be made on both classes and properties, disjointness statements can be made only on classes, and equality/inequality can be asserted between individuals (Horrocks et al. 2003). Names in OWL are international resource identifiers (IRIs) (Duerst and Suignard 2005). There are various syntaxes available for OWL, which serve various purposes. There are tools that can translate between the different syntaxes for OWL.

The following gives a *"ResidentialBuilding"* OWL ontology example, which is written using some of the introduced language constructs. This example shows that a *"ResidentialBuilding"* is a subClass of *"Building"* and has property *"LocatedAt"*.

```
<owl: Class rdf:ID= "ResidentialBuilding">

    <owl: rdfs:subClassOf  rdf:parsetype="Building"/>

    <owl:Restriction>

        <owl:onProperty rdf:resource="LocatedAt" />

        <owl:minCardinality rdfs:datatype="&xsd:Integer">

            1

        </owl:minCardinality>

    </owl:Restriction>

</owl:Class>
```

The activity of creating OWL documents is conceptually different from programming due to its declarative nature. OWL is not a programming language; it is a "declarative" language, which describes a state of affairs in a logical way. OWL is not a schema language for syntax conformance. It does not provide methods to describe how a document should be structured syntactically, and there is no way to enforce that a certain piece of information has to be syntactically presented (Hitzler et al. 2009). Although OWL stores information in a similar way as do databases, OWL is not a database framework. However, technically, databases provide a viable backbone in many OWL ontology-oriented systems.

One important advantage of OWL is that it defines data in terms of semantics by building upon the RDF and assigning a specific meaning to certain RDF triples. By providing a semantic interpretation of the data, OWL makes web information more readily accessible for computers to automatically process. OWL contains rules that can perform certain types of runtime automatic reasoning. In fact, OWL is a knowledge representation and computational logic-based language, which is designed to exchange and reason with knowledge about a domain of interest. Supporting a rich set of inferences is one aspect of OWL that distinguishes it from RDF and RDF Schema. Computer programs can reason the knowledge expressed in OWL to verify the consistency of that knowledge or to make implicit knowledge explicit. Reasoners can be used to infer further information from the current state of affairs of OWL. Thus OWL allows computers to automatically understand structures and meanings of diverse information sources and conduct automatic knowledge inferring or reasoning from existing data and documents. Using OWL for ontology and data integration, it is possible to define and automate the use of ontologies in data integration domains, which are made up of thousands of systems with their own semantic meanings. By binding these diverse systems together in a common ontology through defining a common semantic meaning of data, OWL supports spatial information interoperability at the semantic level.

Although OWL aims to capture knowledge, it does not include all aspects of human knowledge. In fact, it only includes certain parts of human knowledge. In addition, while some of the inferences are quite obvious, others supported by OWL are quite complex, requiring reasoning by cases and following chains of properties. For example, some aspects of RDF, such as the use of classes as instances, interact with other aspects of OWL to create computational difficulties. This may make OWL go beyond the ranges of practical algorithms and implemented reasoning systems. Therefore, solutions for these computational issues need to be provided to increase compatibility with RDF and RDF Schema.

3.3   From UML (Unified Modeling Language) to OWL

3.3.1   What is UML and Why from UML to OWL?

People have begun to actively build ontologies for different applications, for example, an Ontology for Transportation Networks (OTN) developed based on

GDF: http://www.pms.ifi.lmu.de/rewerse-wga1/otn/OTN.owl). Building ontologies is not an easy task. Currently ontologies are typically built by a small number
of people, in most cases researchers, using ontology tools and editors such as
Protégé (Hamilton and Miles 2006). These ontology tools and editors supporting
ontological modeling have improved over the last few years. Many functions are
available now, such as ontology consistency checking, import of existing ontologies, and visualization of ontologies. However, ontology building manually has
proven to be a very difficult and error-prone task and has become the bottleneck
of ontology acquiring processes. For instance, it is unrealistic for non-domain-
experts to use these tools to build high quality ontologies. The question faced by
application practitioners is how to develop ontologies using well-designed and
user-friendly software programs in order to use them in real world applications.
Converting from the Unified Modeling Language (UML) to OWL may be a way
to develop high quality ontologies.

In the standardization approaches, the Unified Modeling Language (UML) has
been identified as a way of providing standard modeling for data. As a standard
modeling language, UML has received wide attention in many fields and were
used as a graphical paradigm to assist human comprehension of concepts and their
relationships. The UML data model supports the exchange of spatial and temporal data and fosters improvements in common spatial data infrastructure through
enhanced data sharing. Further, as an object-oriented data model, the UML data
model enhances the efficiency of database maintenance and increases the performance of applications. Based on the object-oriented (OO) design concept, UML
was developed to help in describing and designing software systems using graphical notations so that the design and viability of a system can be easily understood
(Pilone and Pitman 2005; FGDC 2006). UML has become a widely accepted and
supported standard modeling language in data modeling. Due to its graphic and
symbolic representation of information and relationships, UML makes it easier to
design, modify, and maintain data change over time.

The basic concepts of UML are *class*, *class relationships*, *object instances*,
and *package*. Figure 3.3 shows an example of modeling the "*Pattern*" class, its
attributes and its relationships with the "*TransitRoute*" class and the "*Trip*" class
in transit network data using UML. As shown in Fig. 3.3, transportation features
Pattern, *TransitRoute* and *Trip* can be modeled as "*classes*" in UML. A *class*
represents a group of things that have common states and behaviors, and it is denoted by a rectangular box divided into compartments. Usually the first compart-
ment is used for writing the name of the class, the second holds attributes, and
the third shows operations. The *Pattern* class contains four attributes (*pattern-
Type*, *routeDirection*, *transitServiceTyp*, and *timetableVersion*). At a minimum,
an attribute usually contains a *visibility* property, a *name*, and a *type*, as shown in
Fig. 3.3. In addition, attributes may contain other properties such as *multiplicity*,
uniqueness, and *ordering*. The relationships or connections between transportation concepts or entities can be modeled by the relationships between classes in
UML, which include five different major types of relationships (*Dependency*,
Association, *Aggregation*, *Composition*, and *Generalization*). Attributes can

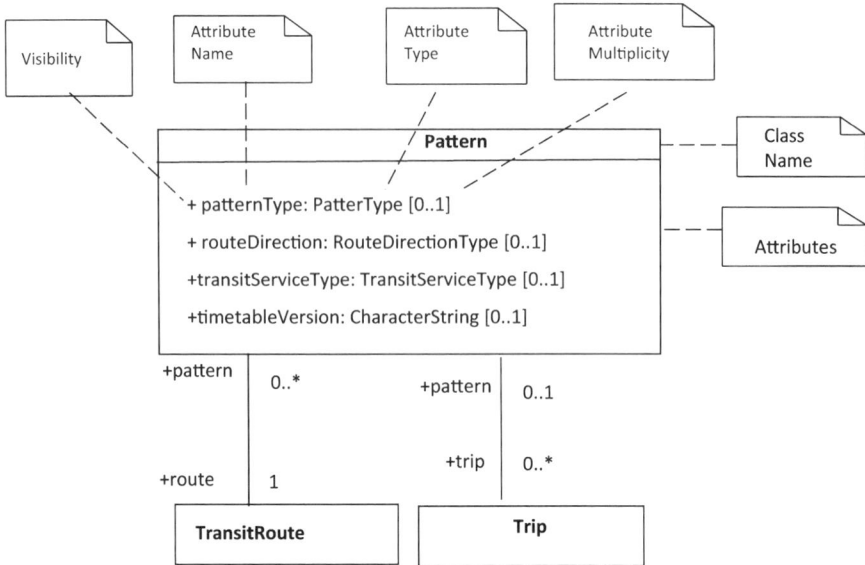

Fig. 3.3 Modeling the *Pattern* class, its attributes and its relationships with the *TransitRoute* class and the *Trip* class in transit network data using UML (Adapted from FGDC 2006)

have simple primitive types (such as *integers, floating-point numbers*, and *CharacterString*), or class types. For example, Fig. 3.3 shows association relationships among *Pattern, TransitRoute*, and *Trip*. In the example of the association relationship between *Pattern* and *TransitRoute*, one pattern only has one route associated with it while one route may be associated with more than one pattern. In the example of the association relationship between *Pattern* and *Trip*, one trip can only be associated with at most one pattern while one pattern may be associated with more than one trip. A specific instance of transportation features is modeled by object instances of object diagrams in UML. Links can be used to tie object instances together.

Classes themselves cannot provide much insight into how a system is designed. The relationships between classes represent connections between concepts or entities. There are several ways for representing relationships between classes in UML. According to the strength of a class relationship, the five different major types of relationships are *Dependency, Association, Aggregation, Composition*, and *Inheritance* (*Generalization*). *Association* is the most common relationship in UML and has explicit notation to express *navigability* and *multiplicity*. *Navigability* means navigating from one class to another, which can be shown by an arrow in the direction of the class to be navigated to. An association may be bidirectional or unidirectional. It is a common practice to omit the arrows if it is bidirectional. A *multiplicity* expression at an end of an association specifies how many objects of the class at the end may take part in that relationship with a single object of the class at the other end of the association. If the association between two elements is complex, an association class can be used to represent the connection.

The specific instance of a feature is modeled by object instances of an object diagram in UML. The *object instance* notation is similar to the *class* notation and an *object* is shown with a rectangle. To differentiate an instance of a class from the class itself, the title of the object instance is underlined. *Links* can be used to tie object instances together. *Links* between two object instances show that the two object instances can communicate with each other. However, to create a link between two object instances, there must be a corresponding association between the classes.

Related UML elements can be grouped into *packages*, which provide a great way to visualize dependencies between parts of a system. A *package* can be expressed using a rectangle with a tab attached to the top left. Elements within the package can be drawn inside the large rectangle. One package can be imported to another package, thus elements in the package can be accessed by another package. In addition, different packages can be merged together by creating relationships between classes of the same name.

In general, the basic concepts or features of spatial data can be modeled by classes in UML. The relationships or connections between concepts or entities can be modeled by the relationships between classes in UML. Specific instances of spatial features are modeled by object instances of object diagrams in UML. Links can be used to tie object instances together.

Through the feature classes and relationships, UML provides a good mechanism to organize and standardize the spatial data. It can capture and express concepts and relationships graphically and make feature relationships intuitively easy to understand. But it does not define the data in terms of semantics. It does not contain rules and formal programming logic that can perform certain types of runtime automated reasoning. Thus, while the UML data model can facilitate human understanding of the structures and meanings of spatial data, it is still difficult for computers to automatically understand and fuse the structures and meanings of diverse information sources. In order to share data from different sources, the data sources have to exactly follow the standard model and terminology. This is obviously time-consuming and demotes the motivation of data sharing, because different applications have to re-do their data model and adjust the data elements. A better way is to have a mechanism to allow computers to access structured collections of information and sets of inference rules to conduct automatic reasoning and data conversion and linkages. This calls for knowledge representation of spatial data using OWL ontology. Hence, converting the Unified Modeling Language (UML) to OWL may facilitate automatic reasoning and data sharing at the semantic level.

There are commonalities between UML and OWL in handling concepts and relationships. Thus it is possible to develop ontologies in OWL through transforming the existing UML models for different applications. In fact, the two representations share a set of core functionalities. For example, both UML and OWL representations define *classes*, *class relationships*, and *relationship cardinalities*. Table 3.3 summarizes the common features between UML and OWL. This provides a foundation to

Table 3.3 Common features between UML and OWL (Based on IBM (2006))

UML features	OWL features
Class, property ownedAttribute, type	Class
Instance	Individual
ownedAttribute, binary association	Property
Subclass, generalization	Subclass subproperty
N-ary association, association class	Class, property
Enumeration	Oneof
Navigable, non-navigable	Domain, range
Disjoint, cover	disjointWith, unionOf
Multiplicity	minCardinality maxCardinality inverseOf
Derived	No equivalent
Package	Ontology
Dependency	Reserved name RDF:properties

develop a direct linkage and transformation mechanism between UML and OWL ontology. The rules and algorithms for the transformation from UML to OWL ontology knowledge base are introduced in the following section.

3.3.2   Rules and Algorithms for Transformation from UML to OWL Ontology Knowledge Base

In this section, the basic rules to be used in the algorithms to transform UML to OWL ontology knowledge base are first introduced, and then a detailed description of the transformation process is given.

Basic Rules

A set of UML class diagrams G is defined as a tuple (V, E), where V is a set of UML classes, and E is the union of a set of binary associations AS and a set of aggregation relations AG.

Suppose that owl is a function that maps a UML class to its translation in OWL.

a. Given $G = (V, E)$, and for any UML class v in V, letting $v = (n, s, p, AT)$, where n is the name of the class, s is the stereotype of the class, p is v's parent classes, and AT is a set of owned attributes, then an OWL class $owl(v)$ with the name n can be created. Also, let $owl(v)$ be the subclass of $owl(p)$, and assume m is the ID of $owl(p)$.

<owl:Class rdf:ID="n">

<rdfs:subClassOf rdf:resource="#m">

</owl:Class>

b. If s is not <<union>>, for each attribute a in AT, let a=(m, t), where m is its name
    and t is the type. Note that t can be either a simple data type, such as string and
    integer, or a complex object data type, such as another UML class *v*.
If *a=(m, t)* and *t* is a built-in simple data type, then a *Datatype* property of the name
mn with *owl(v)* as the domain and *t* as the range is created.

    <owl:DatatypeProperty rdf:ID="mn">

      <rdfs:domain rdf:resource="#n"/>

      <rdfs:range rdf:resource="#t"/>

      <rdfs:subPropertyOf rdf:resource="#dataTypeAttribute"/>

    </owl:DatatypeProperty>

    <owl:DatatypeProperty rdf:ID="dataTypeAttribute"/>

Note that *mn* is made as a sub-property of a distinct property *dataTypeAttribute* in
order to distinguish it from properties translated from associations and aggrega-
tions. The *objectAttribute* below is similar.

    If *t* is a complex object data type such as a UML class, then an object property
of the same name, domain but with *owl(t)* as the range is created. Suppose that *k* is
the ID of *owl(t)*.

    <owl:ObjectProperty rdf:ID="mn">

      <rdfs:domain rdf:resource="#n"/>

      <rdfs:range rdf:resource="#k"/>

      <rdfs:subPropertyOf rdf:resource="#objectAttribute"/>

    </owl:ObjectProperty>

    <owl:ObjectProperty rdf:ID="objectAttribute"/>

If *s* is <<*union*>>, then it is assumed that all attributes in *AT* have UML classes as types and for each *a* in *AT*, if $a = (m, t)$, then let *owl(v)* be a subclass of *owl(t)*.

c. For each relation e in E, let $e = (d1, d2)$, where d1, d2 are two ends of the relation, $d1 = (v1, m1, l1, u1)$ and $d2 = (v2, m2, l2, u2)$, and for each end of the relation $(i = 1,2)$, vi is the UML class, *mi* is the role name, and *li* is the lower bound and *ui* is the upper bound of cardinality. Suppose the name of *v1* is *n1* and the name of *v2* is *n2*. An object property is defined with the name *m2n1*, the multiplicity constraint [*l2, u2*], the domain of *n1* and the range of *n2*.

<owl:ObjectProperty rdf:ID="m2n1">

  <rdfs:domain rdf:resource="#n1"/>

  <rdfs:range rdf:resource="#n2"/>

  <rdfs:subPropertyOf rdf:resource="#association"/>

</owl:ObjectProperty>

The *min* cardinality constraint is specified below, while the *max* constraint is similar.

<owl:Class rdf:about="#n1">

 <rdfs:subClassOf>

  <owl:Restriction>

   <owl:onProperty rdf:resource="#m2n1"/>

   <owl:minCardinality rdf:datatype="&xsd;nonNegativeInteger">

   </owl:minCardinality>

  </owl:Restriction>

 </rdfs:subClassOf>

</owl:Class>

A symmetric object property is also defined with name *m1n2*, cardinality [*l1, u1*], domain of *n2*, and range of *n1*. The *m1n2* can be specified to be the inverse property of *m2n1* as shown below.

<owl:ObjectProperty rdf:ID="m1n2">

<owl:inverseOf rdf:resource="#m2n1"/>

</owl:ObjectProperty>

If *e* is in the set *AS*, then this property is a sub-property of a distinct property association.

<owl:ObjectProperty rdf:ID="association"/>

Similarly, if *e* is in the set *AG*, then this property is a sub-property of a distinct property aggregation. It is assumed that an individual of the domain class is an aggregation of the individuals of the range class. Also, if *e* is an ordered association or aggregation, then this property is a sub-property of a distinct property ordered. Note that it is not needed to specify the lower or upper bound of cardinality if the bound is zero or unlimited. If the lower and upper bounds are the same, it is just needed to specify the cardinality directly.

d. If an UML class has a finite number of instances, the class can be represented as an enumeration class. For example, a FareType class has instances of *full-fare*, *reduced*, *special*, and *transfer*. It can be translated to an OWL class as below.

<owl:Class rdf:ID="FareType>

<owl:oneOf rdf:parseType="Collection">

<owl:Thing rdf:about="#full-fare/>

<owl:Thing rdf:about="#reduced"/>

<owl:Thing rdf:about="#special"/>

<owl:Thing rdf:about="#transfer"/>

</owl:oneOf>

</owl:Class>

e. Spatial and temporal classes can also be defined to distinguish classes with spatial properties from classes with temporal properties.

```
<owl:Class rdf:ID="Spatial">

  <owl:equivalentClass>

  <owl:Restriction>

<owl:onProperty rdf:resource="#geometry"/>

<owl:allValuesFrom rdf:resource="#GM_Object"/>

  </owl:Restriction>

  </owl:equivalentClass>

  </owl:Class>


  <owl:Class rdf:ID="Temporal">

   <owl:equivalentClass>

    <owl:Restriction>

   <owl:onProperty rdf:resource="#time"/>

   <owl:allValuesFrom rdf:resource="#Time"/>

   </owl:Restriction>

   </owl:equivalentClass>

  </owl:Class>
```

With ontology reasoning, any OWL class with object property *geometry* that has the range of *GM_Object* is a subclass of *Spatial*. Any OWL class with object property *time* that has range of *Time* is a subclass of *Temporal*. The OWL codes of *Point* and *Line* subclass of *Spatial* class are written as below:

```
<owl:Class rdf:ID="Point">

    <rdfs: subClassOf rdf:resource="Spatial">

      <owl:Restriction>

    <owl:onProperty rdf:resource="#geometry"/>

    <owl:allValuesFrom rdf:resource="#GM_Point"/>

       </owl:Restriction>

   </owl:Class>


<owl:Class rdf:ID="Line">

    <rdfs: subClassOf rdf:resource="Spatial">

      <owl:Restriction>

    <owl:onProperty rdf:resource="#geometry"/>

    <owl:someValuesFrom rdf:resource="#GM_Curve"/>

       <owl:someValuesFrom rdf:resource="#GM_MultiCurve"/>

       </owl:Restriction>

   </owl:Class>
```

Transformation Algorithms

Based on the above rules, the algorithm to transform UML into OWL ontology knowledge base was developed by Zhang et al. (2008). The transforming process to be used in the algorithm is shown in Fig. 3.4. First, open the XML Metadata Interchange (XMI) file of UML diagrams to check the correctness of the syntax. If the file passes syntax check, the system performs transformation in the following steps:

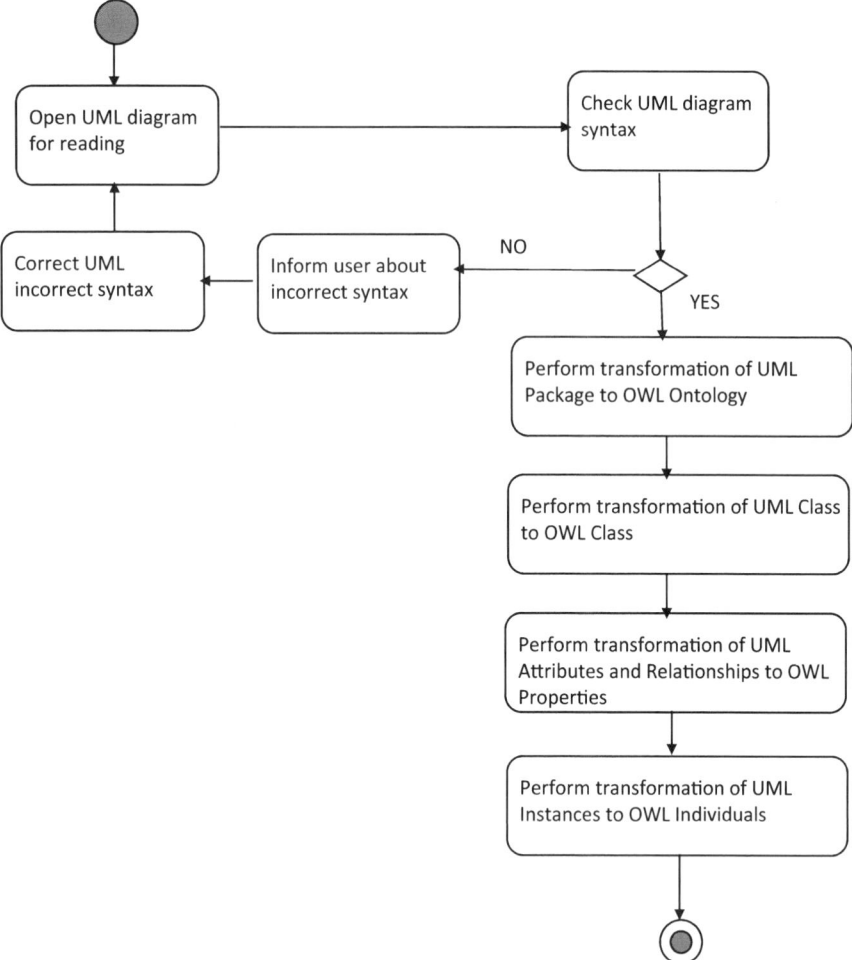

Fig. 3.4 Algorithm for transforming UML into OWL ontology

(a) transform the UML package to OWL ontology;
(b) transform UML classes to OWL classes;
(c) transform attributes of classes and relationships between classes in UML into
      properties in OWL; and finally
(d) transform UML instances into OWL individuals.

Figures 3.5, 3.6, 3.7, and 3.8 show the detailed steps of the transformation process
based on the above described algorithm.

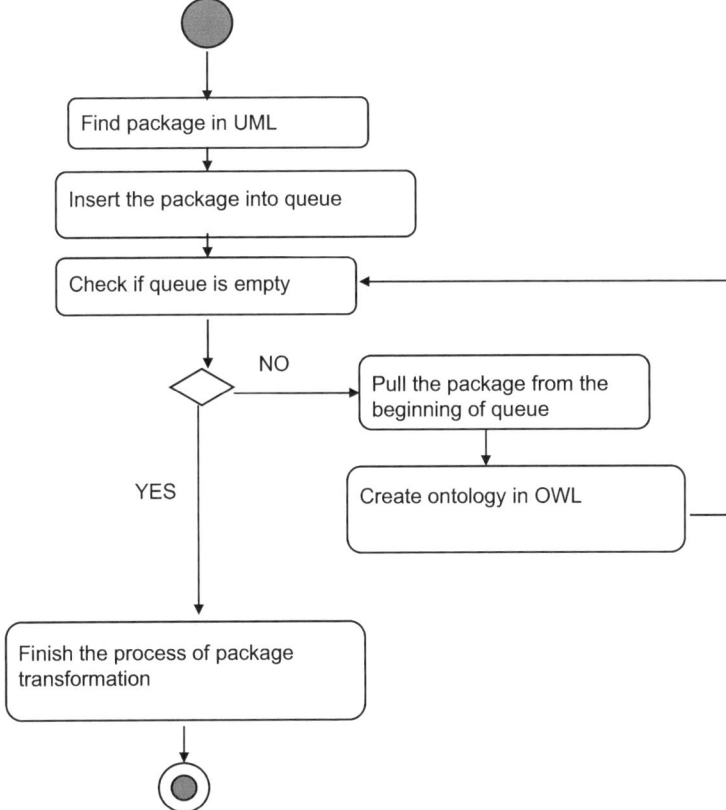

Fig. 3.5 Transformation of UML packages into OWL ontologies

Since both UML and OWL support a module structure, which is called *package* in UML and *ontology* in OWL, the translation of package to ontology is straightforward, as shown in Fig. 3.5. Considering there may not be many packages for an application, a simple linear search algorithm is used to translate packages in UML to OWL ontologies. It operates by checking all elements in UML files one at a time in sequence until a package is found. Once a package is found, it is inserted into a queue. Then a package is pulled out from the beginning of the queue, and translated into OWL ontology. The algorithm processes one package at a time in sequence until the queue is empty.

Similarly, the translation of classes (or subclasses) from UML to classes (or subclasses) in OWL is also simple, as shown in Fig. 3.6. *Classes* in UML can be translated into *classes* in OWL (*owl:Class*) and the *generalization* relationship in UML

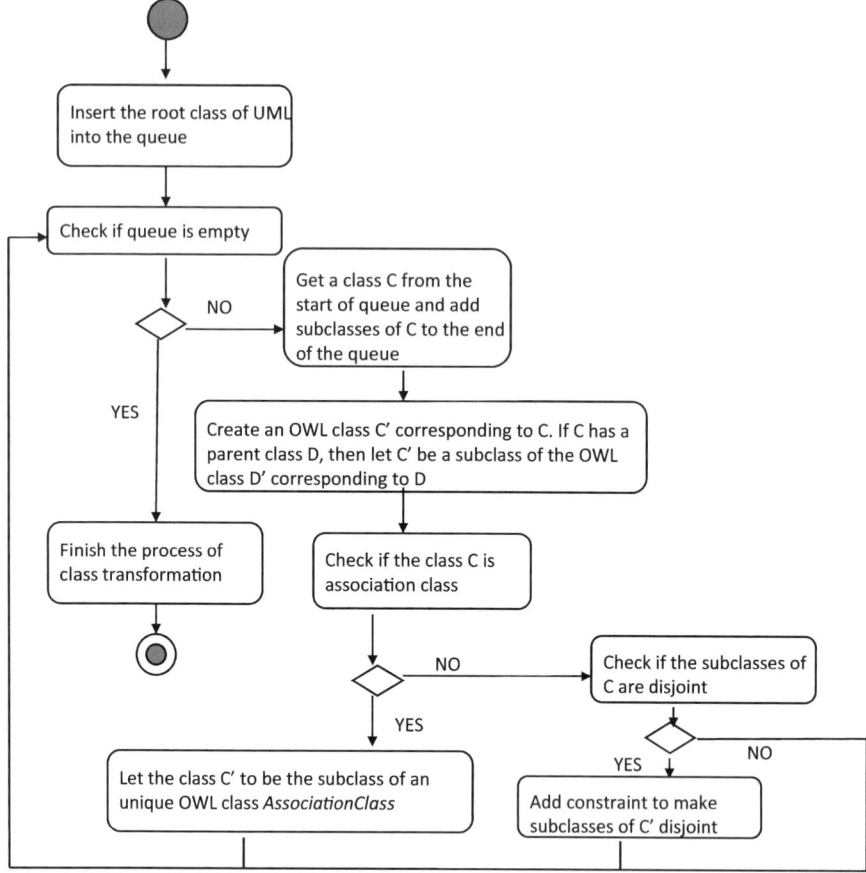

Fig. 3.6 Transformation of UML classes into OWL classes

can be translated to *subclass* in OWL (OWL *rdfs:subClassOf*). The translation uses the breadth-first search algorithm that goes across one hierarchical level of the UML class tree. Root classes are parsed first, then their subclasses, and so on. Because the algorithm uses breadth-first search, it is guaranteed that the parent class in the hierarchy has already been created when some subclasses are being created. One thing that should be checked during the transformation process is whether the class is an association class in UML. If the class is an association class in UML, its corresponding OWL class will be a subclass of a unique OWL class "*AssociationClass*". Another thing that should be checked is whether the subclasses are disjointed. If they are disjointed, those corresponding OWL subclasses should be made disjointed as well.

The process of transforming the UML class owned attributes and class association relationships to *owl:ObjectProperty* and *owl:DatatypeProperty* is shown in Fig. 3.7. If the type of the attribute is a UML *Class*, the UML *ownedAttribute* will be translated to *owl:ObjectProperty*. Otherwise it will be translated to *owl:DatatypeProperty*. The relations among classes in UML are represented by *owl:ObjectProperty* in

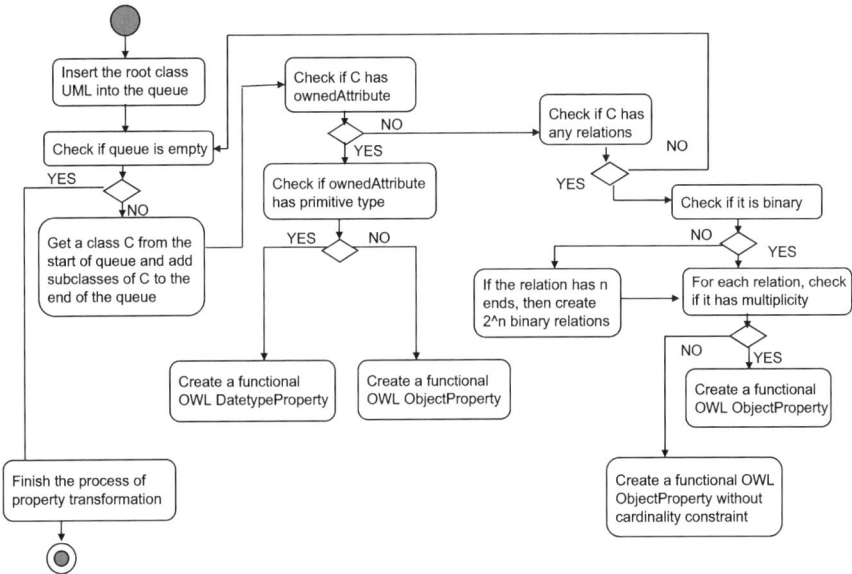

Fig. 3.7 Transformation of UML class attributes and relationships into OWL properties

OWL. OWL accomplishes representation of relations through its range and domain. A binary UML association is translated directly to an *owl:ObjectProperty*. Since the associations in UML are always between types, the OWL property always has its domain and range specified. If a UML binary association has one navigable end and one non-navigable end, it will be translated into a property whose domain is the navigable end and whose range is the non-navigable end. If a UML binary association has two navigable ends, it will be translated into a pair of OWL properties, where one is *inverseOf* the other. If UML associations are not binary and are N-ary associations, the N-ary UML associations will be converted to 2^n binary associations and then translated to OWL object properties.

The multiplicity of UML relations can be translated into cardinality restrictions on the OWL property by giving the minimum (*minCardinality*), maximum (*maxCardinality*), or just an exact cardinality. If a binary UML association has a multiplicity on a navigable end, the translated corresponding OWL property will have the same multiplicity. If a binary UML association has a multiplicity on its both ends, the corresponding OWL property will be an inverse pair each having one of the multiplicity declarations. For an N-ary UML association, any multiplicity associated with one of its UML properties would be applied to the corresponding OWL property.

Given that both UML and OWL support sub-properties (UML generalization of association), the same rules used to transform UML class attributes and relationships into OWL properties will be applied to transform UML subclasses and generalizations of associations into OWL subproperties. To guarantee that the parent properties in the hierarchy have already been created when some subproperties are being created, the breadth-first search algorithm is used in the property and subproperty transformation process.

Fig. 3.8 Transformation of
UML instances into OWL
individuals

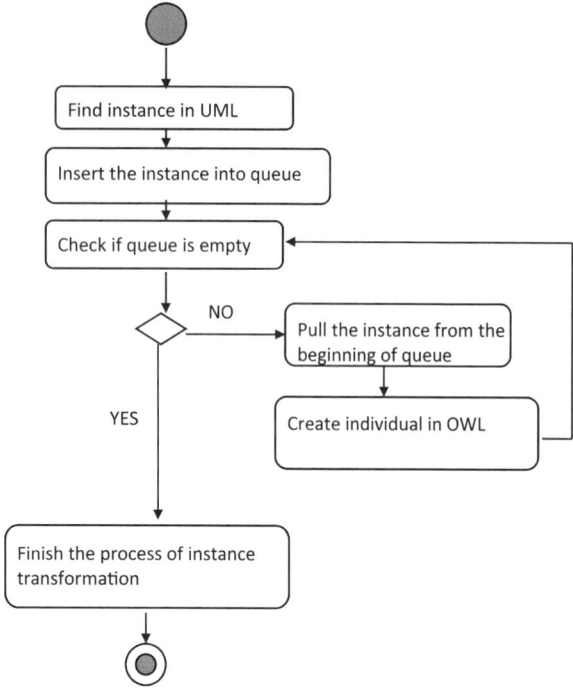

As both languages support instances or individuals, the translation of UML instances into OWL individuals is easy. As shown in Fig. 3.8. The simple linear search
algorithm can be used to check and translate every UML instance in sequence into
an OWL individual.

There are some features which can only be expressed in OWL and some other
features which can only be expressed in UML. These incompatibilities prevent
some elements in UML from being represented in OWL and also cause some OWL
elements unavailable from UML. Some elements in UML that do not have their
counterparts in OWL are provided as follows (IBM 2006):

1. UML permits data to be modeled using names as variables. However, OWL does
   not support variables at all.
2. UML supports behavioral features and classes may have operations. However,
   OWL does not provide the dynamic functions.
3. UML supports complex objects and permits various types of the part-of relationship between classes. However, OWL does not differentiate *Dependency*,
   Association, *Aggregation,* and *Composition* relationships between classes.
4. UML supports access controls and permits a property to be designated read-only
   and also allows classes to have public and private elements. However, OWL
   does not provide these kinds of access controls.
5. UML has *keywords* to be used to extend the functionality of the basic diagrams.
   However, OWL does not offer the *keyword* function to reduce the amount of
   symbols. In addition, UML has abstract class, which typically cannot be instantiated. But there is no OWL equivalent for this.

There are also some features that only exist in OWL (IBM 2006). First, OWL permits a class to be defined as the set of individuals which satisfy a Boolean combination of other classes such as *intersectionOf, unionOf, complementOf.* UML class diagrams have no counterparts. Based on property restrictions, OWL also allows classes to be defined using restriction expressions such as *EquivalentClass.* It allows a class property value to be restricted by a certain value (*hasValue*) and two properties to be stated as equivalent (*equivalentProperty*). However, UML class diagrams do not provide these methods to allow computers to make reliable inferences from involving vague concepts. While OWL can be treated as a sort of predicate definition language, UML provides but does not mandate the predicate definition language OCL (Object Constraint Language). Second, names in OWL do not by default satisfy the unique name assumption. In OWL, the same name always refers to the same object, but different names may also refer to the same object by using *sameAs* and one name can be declared to refer to something different by *differentFrom*. However, names in UML usually by default satisfy the unique name assumption although this is not mandated. Different names usually refer to different objects and cannot be used to refer to the same object. Third, OWL has special properties such as transitive and symmetric; however, UML class diagrams have no corresponding features.

Because of the differences between UML and OWL, UML can only to be used in initial phases of ontology development. These limitations may be overcome using UML extensions. With UML extension notations, UML profiles can support modelers to develop vocabularies and richer ontologies in OWL. Although there is still no final solution to build complex ontologies in OWL using UML, current development of formal semantics for UML is an active area of research.

Although the transformation approach shows a promising way to develop ontologies, there are still some issues yet not to be resolved due to the differences between UML and OWL. Future research should concentrate on developing mechanisms to further reduce the dissimilarity between UML and OWL and take advantage of the merits of both.

3.4   Ontology-Based Reasoning and Rule-Based Knowledge Inference

3.4.1   Using DL (Description Logics) to Represent Knowledge

Because OWL is based on Description Logics (DL), a DL-based reasoner and inference rules are used to collect a knowledge base for automatic service queries on Geospatial Semantic Web. Description Logics is a well-known family of knowledge representation formalisms. They are based on the notion of concepts (unary predicates, classes) and roles (binary relations), and are mainly characterized by constructors that allow complex concepts and roles to be built from atomic ones (Horrocks et al. 1999). The main benefit from DL is that it can solve the subsumption (subconcept/superconcept) and satisfiability (consistency) problems that often exist in the presenting data. A DL reasoner can check whether two concepts equal, satisfy (consist) or subsume each other (Horrocks and Patel-Schneider 1998).

Suppose the letters A and B are used for atomic concepts, the letter R for atomic roles, and the letters C and D for concept descriptions, the main abstract DL syntax notations are listed as followings:

$C, D \rightarrow A$	Atomic concept
T	Universal concept
$\perp$	Bottom concept
$\neg A$	Atomic negation
$C \cap D$	Concept conjunction
$C \cup D$	Concept disjunction
$\forall R.C$	Value restriction
$\exists R.T$	Existential quantification
$\geq nR$	At_most cardinality restriction
$\leq nR$	At_least cardinality restriction
$= nR$	Exactly cardinality restriction

For example, using the above syntax notations, those routes that have multiple lanes can be denoted as *Route* $\cap$ $\exists$ *hasLane*. T and those routes whose lanes are wide can be denoted as *Route* $\cap$ $\forall$ *hasLane.Wide*. Readers are referred to Baader et al. (2003) for further readings about DL.

To provide means for dealing with spatial descriptions used in OGC WFS operator descriptions, such as *equals, disjoint, intersects, touches, contains, crosses*, an extended DL formalism to reason about spatial relations has to be developed (Zhang et al. 2010a). The extended DL provides modeling constructs, which can be used to represent spatial relations as defined roles. The popular and well-known set of the RCC8 (Region Connection Calculus 8) spatial relations for regions (polygons) are adopted in the extended DL formalism (Randell et al. 1992). The RCC (Region Connection Calculus) can be used for qualitative spatial representation and reasoning. RCC8 has 8 basic relations for two regions (polygons). Figure 3.9 shows these eight basic relations for two regions (polygons) in RCC8: (a) DC (disconnected); (b) EC (externally connected); (c) TPP (tangential proper part); (d) NTPP (non-tangential proper part); (e) PO (partially overlapping); (f) EQ (equal); (g) TPPi (tangential proper part inverse); (h) NTPPi (non-tangential proper part inverse). From these basic relations, combinations can be developed. For example, proper part (PP) is the union of TPP and NTPP. Reasoning about spatial configurations among regions (polygons) can be made using the RCC8.

Because most spatial data involve the spatial relations between points and lines, the polygon RCC8 relations are extended to points and lines. Figure 3.10 shows the extended Feature relations among points, lines, and polygons. There are three relations between line and line: Equal (EQ), Disconnected (DC), and Cross (CR); two relations between point and line: Un-Touch (UT) and Touch (TO); three relations between point and polygon: Inside (IN), Tangential Part (TP), and Outside (OUT); four relations between line and polygon: Inside (IN), Tangential Part (TP); Cross

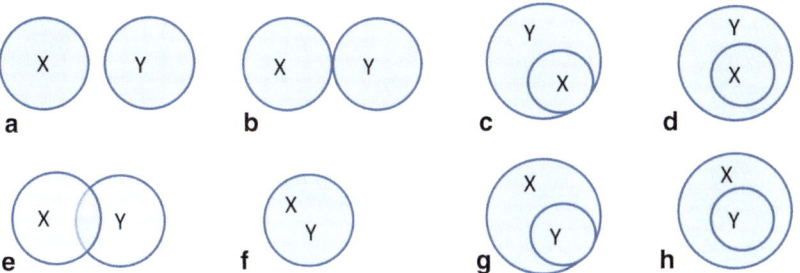

Fig. 3.9 The basic relations for two regions (polygons) in RCC8: (**a**) DC (disconnected); (**b**) EC (externally connected); (**c**) TPP (tangential proper part); (**d**) NTPP (non-tangential proper part); (**e**) PO (partially overlapping); (**f**) EQ (equal); (**g**) TPPi (tangential proper part inverse); (**h**) NTPPi (non-tangential proper part inverse)

(CR) and Disconnected (DC). Using these base spatial relations, indefinite spatial knowledge can be expressed as a union of different possible base spatial relations.

The reasoning tasks depend on the extended DL formalism for representing spatial knowledge. Spatial reasoning can be done by deriving spatial relationships from given knowledge such as the existing transit network maps. Given, for instance, the following spatial descriptions - "Bus stop A touches Route B" and "Route B is located in Waukesha", software programs can automatically derive that bus stop A is also located in Waukesha based on the extended DL formalism.

To query the individual spatial entities, a spatial relation is computed from the geometry of the exiting maps, for example, the transit network maps including bus stop, bus route, and street. The spatial relation is represented by means of role assertions in ABox, such as, (i,j): *TPPI*, (i,k): *CR*. For example, consider the instance retrieval query *Public_route_crossing_a_river (?x)* on ABox:

$$A = \left\{ i : \text{route} \cap \text{public}, k : \text{river}, (i,k) : CR \right\}$$

In order to retrieve the instances of *Public_route_crossing_a_river,* each individual instance is considered and checked separately. Let's consider i. Verifying whether i is an instance of *Public_route_crossing_a_river* is reduced to checking the unsatisfiability of:

$$A \cup \left\{ i : \neg \text{Public\_route\_crossing\_a\_river} \right\} \text{ or}$$
$$A \cup \left\{ i : \left(\neg \text{route} \cup \neg \text{public} \cup (\forall CR. \neg \text{river}) \right) \right\}$$

This ABox is unsatisfiable; thus, i is a *Public_route_crossing_a_river.*

In addition, inference rules can be defined to enable further query from the OWL knowledge base. The inference rules can be used to derive object properties based on the *datatype* properties. To facilitate the ontology query, the inference rules can also be defined to incorporate the use of semantic constraints such as subclass and

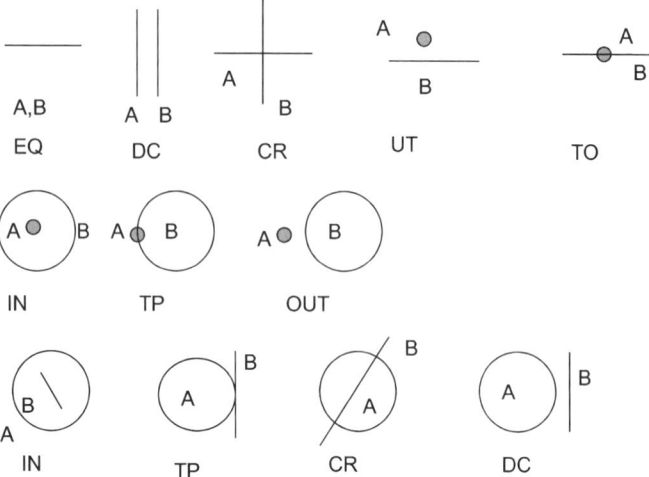

Fig. 3.10 Feature relations. Line and Line relations: EQ=Equal, DC=Disconnected, CR=Cross. Point and Line relations: UT=Un-Touch, TO=Touch. Point and Polygon relations: IN=Inside, TP=Tangential Part, OUT=Outside. Line and Polygon relations: IN=Inside, TP=Tangential Part, CR=Cross, DC=Disconnected

subproperty relations and the domain and range restrictions. An inference rule i written in Datalog-like notations has two parts: the head written as $i.head$ and the body written as $i.body$. The head of the rule has the form of a RDF triple while the body is a set of RDF triples plus some filters.

Definition 1 An inference rule has the form of $r(X){:}-R(Y, Z), F(Z)$, where r is a RDF triple, R is a set of RDF triples, F is an optional set of filters, and X, Y, Z are sets of variables. X is a subset of Y and they contain variables referring to RDF instances. Z is a set of variables referring to datatype values or geometries.

Table 3.4 shows two examples of inference rules. Rule $I1$ says that Property *route_stop* relates a route to a link if the route ID matches the route ID of a pattern, the pattern's ID matches the pattern ID of a link-sequence, and the link-sequence's link ID matches the link's ID. Rule $I2$ says that a stop one has a nearby stop two if the stop two's geometry is within a distance of 1 unit from the stop one's geometry. Note that the filter in Rule $I2$ is similar to the OGC filters. Here the unit of distance is omitted. See Zhao et al. (2008) for more information about the inference rules.

3.4.2 *Knowledge Reasoning Based on Description Logic and Inference Rules*

Because OWL is based on a Description Logic (DL), a DL reasoner can be used to compare (semantically) descriptions written in OWL and automatically reaping the wealth of semantic information in the OWL ontologies that describe relations

Table 3.4 Two examples of inference rules

(*I1*)	(*I2*)
?route route_link ?link :-	?stop nearby_stop ?other :-
?route rdf:type Route,	?stop rdf:type Stop,
?route hasID ?rid,	?other rdf:type Stop,
?pattern rdf:type Pattern,	?stop geometry ?geom,
?pattern hasID ?pid,	?other geometry ?g,
?pattern routeID ?rid,	filter(
?linkseq rdf:type LinkSequence,	DWithin(?g, ?geom, 1)
?linkseq hasID ?lid,	).
?linkseq patternID ?pid,	
?link rdf:type Link	
?link hasID ?lid.	

between ontological concepts, such as *subsumption* (*hyponym-hypernym*) and *equivalence* (*synonym*) relations. For example, by inference sub/super_concepts "*TimePoint*" and "*Stop*" for transportation transit network applications, computers can understand that a "*TimePoint*" must be a "*Stop*" and a "*TimePoint*" inherits all the properties and constraints of a "*Stop*". By allowing the definition of relations between concepts in enhanced semantic web services, it is possible to express statements such as X is part of Y or more generally that a relation R exists between X and Y. This provides a powerful framework for defining and comparing service descriptions to allow matching and integrating semantically heterogeneous spatial data.

One of the characteristics of DL is that it enables systems built on it to infer implicitly represented knowledge from the knowledge that is explicitly contained in the knowledge base. To infer implicitly represented knowledge, the following two axioms are often applied:

Terminological Axioms If C, D are concepts and R, S are roles, then $C \sqsubseteq D$ ($R \sqsubseteq S$) is called an inclusion axiom, which means that concept C (role S) is more specific than concept D (role S). Also, $C \equiv D$ ($R \equiv S$) is called an equivalence axiom, which means C and D (R and S) are equivalent and is an abbreviation of the pair of axioms $C \sqsubseteq D$ and $C \sqsupseteq D$ ($R \sqsubseteq S$ and $S \sqsupseteq R$).

The notation of DL (Baader et al. 2003) used to describe the knowledge base can be directly translated to OWL-DL syntax. Assuming knowledge about a transit system is to be defined, an example is given below to show how to use DL to express the knowledge base for the automatic geospatial feature matching engine.

Example Knowledge "A TimePoint is a TransitStop which is associated with at least one scheduled-time" is defined by the following concept inclusion axiom

$$\text{TimePoint} \sqsubseteq \text{TransitStop} \sqcap \exists\text{associatedWith.Time}$$

where *TransitStop* and *Time* are either primitive or defined concepts, and *associatedWith* is a role. The axiom implicitly defines *TimePoint* as a sub-concept of *TransitStop*.

Using OWL/RDF syntax the above knowledge would be written as:

```
<owl:Class rdf:about="#TimePoint">

   <rdfs:subClassOf>

     <owl:Class>

       <owl:intersectionOf rdf:parseType="Collection">

         <rdf:Description rdf:about="#TransitStop"/>

         <owl:Restriction>

           <owl:onProperty rdf:resource="#associatedWith"/>

           <owl:onClass rdf:resource="#Time"/>


           <owl:minQualifiedCardinality
               rdf:datatype="&xsd;nonNegativeInteger">1
           </owl:minQualifiedCardinality>
         </owl:Restriction>
       </owl:intersectionOf>
     </owl:Class>
   </rdfs:subClassOf>
</owl:Class>
```

Defined concepts ("if and only if") can be added in the knowledge base and exploited to automatically enrich the given basic annotations. Thus, users' own required concepts can be defined, for example, "*public_route_crossing_a_river*" can be defined as "*route which is public and crosses a river*" with a TBox axiom:

$$\left[\text{Public\_route\_crossing\_a\_river} \equiv \text{route} \cap \text{public} \cap \exists \text{cross.river}\right]$$

By providing a formal conceptualization of the domain (defines precisely the concepts and the relationships) and using the aforementioned DL and inference rules, the Geospatial Semantic Web facilitates knowledge sharing and reuse via automatic machine processing. Most basic spatial relations can be geometrically computed and asserted in the ABox of an OWL DL knowledge base. With knowledge reasoning, software can detect implicit information in formal conceptual models for geospatial domain objects and reason the relations of geospatial data; thus the Geospatial Semantic Web can support powerful queries.

3.4.3 Spatial Rules for Implicit Knowledge

Spatial rules can be defined for reasoning over spatial relations between objects in space. These spatial reasoning rules can be used as the deduction rules for automatic derivation of implicit spatial relations (Zhang et al. 2010b). These rules can be defined based on the literature on spatial reasoning (e.g. Sun and Li 2005). For example, the facts that the town of Mansfield is located inside the state of Connecticut and the state of Connecticut is inside the New England region imply that the town of Mansfield is also inside the New England region. Some examples of the rules are given below.

Note that each of the rules below has the form of

conclusion :: condition_1, condition_2, ..., condition_n

The semantics of the rules is that if all the conditions are true, then so is the conclusion.

Rule 1
Transitivity of left (right)_of, above, behind, inside, east, west, north, northwest, northeast, southwest, and southeast. This rule indicates the transitivity of some of the relations. Let x denote any relation in { left(right)_of, above, behind, inside, east, west, north, northwest, northeast, southwest, and southeast }. Each such x has the following rule: A x C:: A x B, B x C.

Rule 2
This rule captures the interaction between the relations involving left-of, above, behind, and the relation involving overlaps. Let x denote any of the relation

symbols- left-of, above and behind. Each such z have the following rule: A x D:: A x B, B overlaps C, C x D.

Rule 3

This rule captures the interaction between the relations involving left-of, above, behind, outside, and the relation involving inside. Let x denote any relation symbol in {left-of, above, behind, outside}. Each such x has the following two rules:

(a) *A x C:: A inside B, B x C*
(b) *A x C:: A x B, C inside B*

Rule (b) is redundant for the case when x is the relation symbol outside; for the other cases (a) and (b) are independent.

Rule 4

Symmetry of equal, overlaps, externally connected, and disjoint: this rule captures the symmetry of equal, overlaps, externally connected, and disjoint. Let x denote any of these relations. Each such x has the following rule: A x B:: B x A.

Rule 5

Inverse Property: This rule says that the following directional relations are inverses of each other:

A North B::B South A
A Northeast B::B Southwest A
A Northwest B::B Southeast A
A East B::B West A
A left B::B right A
A Above B::B Below A

Rule 6

This rule allows one to deduce that two objects are outside each other if one of them is to the left of, or above, or behind the other object. Let x denote any of the relation symbols in {left-of, above, behind}. Each such x has the following rule: A outside B:: A x B

One of the characteristics of these rules is that they enable systems built on them to infer implicitly represented knowledge from explicit knowledge in the knowledge base. These rules can help employ reasoning to support intelligent queries. By defining the above rules, the implicit knowledge of spatial relations can be obtained. The spatial reasoning is necessary because users may like to use the qualitative reasoning more than precision quantitative measurements for query. For example, users are in most cases interested in *whether Mansfield is located inside Connecticut*, instead of *whether Mansfield has smaller latitude than Connecticut*. The flexible query can be done by using the above rules.

3.5   Chapter Summary

This chapter introduces ontology languages RDF and OWL. RDF has come to be used as a general method for conceptual description or modeling of information in the Web using a Graph-based data model. The core structure of the RDF is a set of triples, which are also known as RDF graphs. Each triple consists of three components—a subject, a predicate, and an object. The RDF triples can be visualized by a connected RDF graph, which consists of nodes and arcs. There are three types of nodes in an RDF graph—IRIs, literals, and blank nodes. The OWL Web Ontology Language is an ontology language for the Semantic Web with formally defined meanings. OWL ontologies themselves are primarily exchanged as RDF data and can be used along with RDF. But OWL facilitates greater machine interpretability of data content than RDF does through providing additional vocabularies along with a formal semantics.

OWL ontology building manually has proven to be a very difficult and error-prone task, and becomes the bottleneck of ontology acquiring processes. Converting from the Unified Modeling Language (UML) to OWL may be a way to develop high quality ontologies. There are commonalities between UML and OWL in handling concepts and relationships. Thus it is possible to develop ontologies in OWL through transformation of the existing UML models for different applications. In this chapter, the basic rules to be used in the algorithm to transform UML to OWL ontology knowledge base are introduced, and then a detailed description of the transformation process is given.

Finally, this chapter introduces knowledge about ontology-based reasoning and rule-based knowledge inference. Because OWL is based on Description Logics (DL), a DL-based reasoner and inference rules can be used to collect a knowledge base for automatic service queries on Geospatial Semantic Web. By providing a formal conceptualization of the domain (defines precisely the concepts and the relationships) and using the aforementioned DL and inference rules, the Geospatial Semantic Web facilitates knowledge sharing and reuse via automatic machine processing.

References

Antoniou G, Van Harmelen F (2004) Web ontology language: owl. In: Staab S, Studer R (eds) Handbook on ontologies. Springer, Heidelberg, pp 67–92

Baader F et al (2003) The description logic handbook: theory, implementation, and applications. Cambridge University Press, Cambridge

Bechhofer S et al (2009) Owl: web ontology language. In: Liu L, Özsu MT (eds) Encyclopedia of database systems. Springer, Heidelberg, pp 2008–2009

Brickley D, Guha RV (2014) RDF Schema 1.1. W3C Recommendation, 25 February 2014. http://www.w3.org/TR/2014/REC-rdf-schema-20140225/. The latest published version is available at http://www.w3.org/TR/rdf-schema/. Accessed 15 May 2015

De Vergara JL, Villagrá VA, Berrocal J (2004) Applying the web ontology language to management information definitions. IEEE Commun Mag 42:68–74

Duerst M, Suignard M (2005) RFC 3987: internationalized resource identifiers (IRIs). http://www.
    ietf.org/rfc/rfc3987.txt. Accessed 15 May 2015
FGDC (2006) Information technology-geographic information framework data content standard.
    http://www.fgdc.gov/standards/projects/incits-l1-standards-projects/framework/documents-2/
    transtransit20061115-2017.pdf. Accessed 15 May 2015
Hamilton K, Miles R (2006) Learning UML 2.0. O'Reilly publisher. http://safari.oreilly.
    com/0596009828. Accessed 15 May 2015
Hayes PJ, Patel-Schneider PF (2014) RDF 1.1 Semantics. W3C Recommendation, 25 Feb 2014.
    http://www.w3.org/TR/2014/REC-rdf11-mt-20140225/. The latest edition is available at http://
    www.w3.org/TR/rdf11-mt/. Accessed 15 May 2015
Hitzler P et al (2009) OWL 2 web ontology language primer. W3C recommendation, 27 Oct 2009.
Hitzler P et al (2012) OWL 2 Web Ontology Language. http://www.w3.org/TR/2012/REC-owl2-
    primer-20121211/. Accessed 15 May 2015
Horrocks I, Patel-Schneider PF (1998) Comparing subsumption optimizations. In: Franconi E et al
    (eds) Proceedings of the 1998 description logic workshop (DL'98) (CEUR-WS.org, Electronic
    Workshop Proceedings 11), Povo-Trento, pp 90–94. http://ceur-ws.org/Vol-11/. Accessed 12
    July 1998
Horrocks I et al (1999) Practical reasoning for expressive description logics. In: Ganzinger H et al
    (eds) Proceedings of the 6th international conference on logic for programming and automated
    reasoning (LPAR'99), lecture notes in artificial intelligence, vol 1705. Springer-Verlag, Berlin,
    pp 161–180
Horrocks I, Patel-Schneider PF, Van Harmelen F (2003) From SHIQ and RDF to OWL: the mak-
    ing of a web ontology language. Web Semant: Sci Serv Agents World Wide Web 1: 7–26
IBM (2006) Ontology definition metamodel. http://www.omg.org/docs/ad/06-05-01.pdf. Accessed
    15 June 2007
McGuinness DL, Van Harmelen F (2004) OWL web ontology language overview. W3C recom-
    mendation, 10 March 2004
Motik B et al (2009) Owl 2 web ontology language direct semantics. W3C Recommendation, 27
    Oct 2009
Pilone D, Pitman N (2005) UML 2.0 in a Nutshell. O'Reilly publisher.
Randell DA et al (1992) A spatial logic based on regions and connections. In: Proceedings of the
    International Conference on Principles of Knowledge Representation and Reasoning (KR'92),
    Cambridge, Massachusetts, 26–29 Oct 1992
Schreiber G, Raimond Y (2014) RDF 1.1 primer. http://www.w3.org/TR/2014/NOTE-rdf11-
    primer-20140225/. Accessed 15 May 2015
Sun H, Li W (2005) Spatial reasoning based on rules. In: Mechanisms, symbols, and models
    underlying cognition, lecture notes in computer science, vol 3561. pp 469–480
Van Harmelen F, McGuinness DL (2004) OWL web ontology language overview. World Wide
    Web Consortium (W3C) Recommendation
Zhang C, Peng ZR, Zhao T et al (2008) Transforming transportation data models from unified
    modeling language to web ontology language. Transp Res Rec: J Transp Res Board 2064:81–89
Zhang C, Zhao T, Li W et al (2010a) Towards logic-based geospatial feature discovery and integra-
    tion using web feature service and geospatial semantic web. Inter J Geogr Info Sci 24:903–923
Zhang C, Zhao T, Li W (2010b) Automatic search of geospatial features for disaster and emer-
    gency management. Inter J Appl Earth ObsGeoinfo 12:409–418
Zhao T et al (2008) Ontology-based geospatial data query and integration. In: Geographic infor-
    mation science, lecture notes in computer science, vol 5266. pp 370–392

Chapter 4
Ontology Data Query in Geospatial Semantic Web

4.1 Introduction

Semantic interoperability is a core research topic for integrating, interlinking, and retrieving vast geo-referenced and multi-perspective geospatial knowledge. Geospatial Semantic Web offers the support of semantic interoperability and extends the geospatial infrastructure vision from a data archive and infrastructure to a knowledge engine, which enables more powerful reasoning and information retrieving from heterogeneous and contradictory conceptual models and scientific data in different sources. Geospatial Semantic Web promises better retrieval of geospatial information by explicitly representing the semantics of data through ontologies, which can be understood and processed by computers. It also promotes sharing and reuse of spatial data for a wide variety of applications by using standardized Semantic Web languages such as RDF to encode spatial data.

The downside of representing structured geospatial data in these languages is that it can result in inefficient data access. The GeoSPARQL protocol was proposed by OGC (Open Geospatial Consortium) as an extension of SPARQL for querying geographic RDF data. Runtime cost of GeoSPARQL queries can be dominated by that of spatial join operations due to the fine-grained nature of RDF data model. Lack of spatial indices is one source of the performance problem for GeoSPARQL queries due to inefficient processing of spatial joins. This problem is inherent in the RDF representation of spatial data, which consists of loosely connected data objects related by object properties. Even if spatial objects are originally stored in related database tables, once they are transformed to RDF objects, the spatial indices are lost. It is possible to recreate indices for RDF objects with spatial attributes. However, pre-computing spatial indices does not guarantee performance improvement since the RDF queries are much more flexible than database queries and it is difficult to predict which spatial objects should be indexed and how. It may be more practical to implement extensions to GeoSPARQL query engine to create spatial indices on demand.

C. Zhang et al., *Geospatial Semantic Web*, DOI 10.1007/978-3-319-17801-1_4

Another source of the performance problem for querying a spatial knowledge base is the way by which spatial attributes are stored in RDF data sets. Spatial attributes are usually stored as string literals that conform to certain formats such as WKT or GML. The GeoSPARQL query engine that implements spatial operators and filter functions has to parse these strings to recover the spatial coordinates for spatial computation. A naïve implementation of a spatial operator or a filter function in GeoSPARQL treats its spatial inputs as plain strings and has to parse the strings to retrieve spatial contents such as X and Y coordinates. Repeatedly parsing the spatial inputs imposes a very large runtime overhead.

The lack of indices also prevents adopting distributed query algorithms on a geospatial knowledge base. Since spatial objects are not indexed, GeoSPARQL query engine cannot partition ontology data into subsets to be processed in parallel. As a result, GeoSPARQL query can only be processed as a single-threaded program. Even with pre-computed spatial indices, partitioning spatial ontology data may not improve performance since the targeted data may not be evenly distributed in the indices.

In this chapter, we will first discuss the ontology query protocols for Geospatial Semantic Web, spatial indexing, and spatial join algorithms. Then we introduce a parallel approach for improving the runtime performance of GeoSPARQL engine by separating spatial and non-spatial components of the queries to optimize and parallelize spatial joins. This approach does not pre-compute spatial indices for geospatial ontology, and it also does not implement spatial extensions of a query engine to use the indices. Instead, this approach separates spatial components from non-spatial components in a Geo-SPARQL query, and it processes spatial sub-queries after completely processing non-spatial sub-queries. The main benefit of this approach is that a smaller set of ontology objects can be obtained after non-spatial sub-queries so that their spatial attributes can be cached for subsequent spatial computation including spatial indexing and spatial joins. Since the parsed spatial attributes are cached, the overhead caused by repeatedly parsing spatial literal strings can be avoided. Improved query efficiency can facilitate the access to spatial ontology information for multiple users through highly intensive geo-computation processes over a Geospatial Semantic Web, particularly for time-critical applications such as disaster response.

4.2   SPARQL

SPARQL is a declarative query language for RDF data sets (SPARQL 2008). To extract useful information from RDF data sets, a user can write a query to select data objects from RDF data sets based on the basic graph patterns specified by a

composition of triple patterns and filter expressions. A basic SPARQL *select* query
has the following form:

> **Select** *<a sequence of variables>*
>
> **Where**
>
> *{*
>
>   *<triplepattern>*
>
> **Filter** *(<Boolean expression>)*
>
> *}*

A triple pattern consists of basic triple patterns combined by logical conjunction
(by default), "union", or "optional" operators. To answer a *select* query, a SPARQL
implementation first determines the set of RDF triples in the data sets that match
the specified triple patterns and also satisfy the filter expressions, and then returns
the bindings of the *selected* variables in the matched triples. Each triple pattern has
the form of *subject predicate object*, where *subject* can be a RDF term or a variable,
predicate can be an IRI or a variable, and *object* can be a RDF term or a variable. A
RDF term is either a RDF literal, blank node, or IRI.

Example 4.1 Consider a RDF data set that describes the containment relation of cities and states, as well as their population sizes. To find the cities that have more than
1/10th of populations of corresponding states, we can write the SPARQL query:

> **Select** *?city ?state*
>
> **Where**
>
> *{*
>
>   *?city locatedIn ?state .*
>
>   *?city population ?cp .*
>
>   *?state population ?sp .*
>
>   **Filter***(?cp> ?sp / 10)*
>
> *}*

This query could return the data that includes

City	State
Milwaukee	Wisconsin
Chicago	Illinois

In Example 4.1, the *selected* variables are *?city* and *?state*. The triple patterns and filter expression in the example are implicitly joined by a logical conjunction so that a solution to the query must satisfy all of the triple patterns and the filter expression. A basic triple pattern such as

$$?city\ locatedIn\ ?state$$

specifies that *?city* and *?state* must be instantiated by values that are related by the predicate *locatedIn* in the RDF data set, such as:

Milwaukee *locatedIn* Wisconsin.

Chicago *locatedIn* Illinois.

Moreover, the triple pattern

$$?city\ population\ ?cp$$

finds out the population of *?city* and unifies it with the variable *?cp*. Similarly, the variable *?sp* holds the population of *?state*. Finally, the filter expression

$$\textbf{Filter}\quad (?cp > ?sp / 10)$$

ensures that the population of the *?city* is more than 1/10th of population of the *?state*.

The results of the SPARQL queries may be sorted by some key values. Also, prefix declarations may be added to the query by introducing the prefix shorthand for replacing namespaces in IRIs.

In addition to *select* queries, SPARQL also supports the following forms of queries:

- A *construct* query that returns a RDF graph constructed by instantiating the variables in given triple templates with query results;
- An *ask* query that tests whether a given query has a solution;
- A *describe* query that returns a RDF graph that describes resources.

4.3 GeoSPARQL

GeoSPARQL (OGC11-052r4 2012) is an extension of SPARQL query language for retrieving geospatial information from RDF data sets (Battle and Kolas 2011). The language extensions include a topological ontology for the representation and

Fig. 4.1 Major components of GeoSPARQL

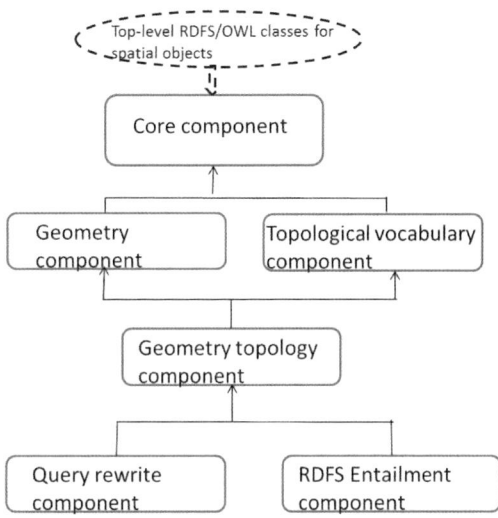

qualitative reasoning of spatial objects and a SPARQL query interface that includes a set of inference rules for transformation query and interpretation, and a set of SPARQL extension functions for quantitative reasoning.

Figure 4.1 illustrates the major components of GeoSPARQL. The *core* component defines top-level RDFS/OWL (RDF Schema/Web Ontology Language) classes for spatial objects. The *Geometry* component describes the geometry vocabulary and non-topological query functions for geometry objects. The *Topological vocabulary* component expresses RDF properties for asserting topological relations between spatial objects. The *Geometry topology* component identifies topological query functions. The *Query rewrite* component defines transformation rules for computing spatial relations between spatial objects based on their associated geometries. Finally, the *RDFS entailment* component introduces a mechanism for matching implicit RDF triples that are derived based on RDF and RDFS semantics.

4.3.1   Topological Representation of Spatial Objects

GeoSPARQL uses GML and WKT literals to describe geometries and features (spatial data objects) using the class *geo:Geometry* and *geo:Feature*, which are related by the properties *geo:hasGeometry* and *geo:hasDefaultGeometry*. The geometry class has a number of properties to specify its dimensions and its serialization. The property *geo:hasSerialization* relates a geometry object to its text-based serialization such as WKT, which is described by the RDFS datatype *geo:wktLiteral*. For example, a WKT literal of the following form represents a point with a default WGS 84 geodetic latitude-longitude spatial reference system:

"*Point*(−83 34)"^^<http://www.net/ont/geosparql#wktLiteral

WKT literals also encode lines and polygons, as well as the geometry collection including multi-points, multi-lines, and multi-polygons. A more verbose form of serialization is GML literals specified by the property *geo:gmlLiteral*. The two sub-properties of *geo:hasSerializations* are *geo:asWKT* and *geo:asGML*, which relate the WKT and the GML serializations to a geometry respectively.

Example 4.2 We can define the city of Milwaukee as a feature object by using the following RDF data:

gn:City	rdfs:subClassOf	geo:Feature.
gn:Milwaukee	gn:name	"Milwaukee";
	rdf:type	gn:City.
gn:Milwaukee	geo:hasGeometry	_p
_p	rdf:type	geo:Polygon
	geo:asWKT	"Polygon(...)"^^geo:WKTLiteral

where *gn* is the prefix of a namespace URI and the coordinates are omitted from the WKT literal.

The primary reason for literal representation of a geometry is that it can be treated as a single unit to be passed to external functions for computation and to be returned from a query. An unfortunate consequence, however, is that the runtime cost of parsing geometry literals can become overwhelmingly expensive for any computation that manipulates a large number of geometries such as a spatial join operation.

4.3.2 Extension Functions for Quantitative Reasoning

GeoSPARQL specifies extension functions for performing non-topological spatial operations such as *distance, buffer, convex hull, intersection, union, difference, symmetric difference, envelop*, and *boundary*. These functions are used in filter expressions.

These extension functions have typing restrictions on their arguments. For example, the *distance* function has the following type declaration:

$$geof:distance \ (geom1: \ ogc:geomLiteral,$$

$$geom2: \ ogc:geomLiteral,$$

$$units: \ xsd:anyURI): \ xsd:double$$

The arguments to the distance function must be geomLiteral type, which may be either a WKT literal or a GML literal. The unit argument can be meter or specified by some URIs. The spatial reference system of the calculation depends on the first parameter. A distance function returns a double value. The result of the function would be invalid if the first two parameters do not use the same spatial reference system.

Other functions may not depend on the unit parameter. For example, the intersection function has the following prototype declaration:

geof:intersection (geom1: ogc:geomLiteral,

geom2: ogc:geomLiteral): ogc:geomLiteral

which takes two geometry parameters and returns a geometry that represents the point at the intersection of the two parameters. The return type does not narrow down to the WKT or GML literal types so that callers to the function may have to dynamically determine the result type.

Example 4.3 We can find the cities that are within 100 kilometers of Milwaukee using the following *distance* function:

Select *?c*

Where {

?c	rdf:type	gn:City ;
	geo:hasGeometry	?g .
?g	geo:asWKT	?w1.
gn:Milwaukee	geo:hasGeometry	?m.
?m	geo:asWKT	?w2.

Filter *(geof:distance(?w2, ?w1, uom:metre) < 100000)*

}

where *uom* is the prefix for the distance unit—*metre*.

4.3.3   Extension Functions for Topological Relations

GeoSPARQL specifies extension functions for topological relations of the three
spatial relation families: Simple Features, Egenhofer, and RCC8. Moreover, it specifies a common function: *geof:relate*, which returns true if its first two geometry
parameters are spatially related based on the third parameter. The third parameter is
a string consisting of T (true), F (false), and * (wildcard) characters that represent a
DE-9IM intersection pattern.

geof:relate (geom1: ogc:geomLiteral,

geom2: ogc:geomLiteral,

pattern-matrix: xsd:String): xsd:boolean

DE-9IM is a model of the spatial relations between two geometry objects. It char-
acterizes the spatial relation between two spatial objects by a 3×3 matrix, each
entry of which is the dimension of the intersection between the interior, boundary,
or exterior region of one object and that of the other object. DE-9IM uses -1 for the
dimension of empty intersection, 0 for points, 1 for lines, and 2 for areas. A simplified form of DE-9IM uses Boolean value F for dimension -1 and T for dimensions
0, 1, or 2. A DE-9IM intersection pattern is a string representation of the intersection
matrix with * as wildcard that matches to either T or F.

The DE-9IM intersection pattern can be written in a string of the structure:

II IB IE BI BB BE EI EB EE.

where each combination *xy* with *x* in {I, B, E} and *y* in {I, B, E} represents the
dimension of the intersection between region x of object 1 and region y of object
2. The character I represents the interior region of a spatial object, B represents the
boundary, and E represents the exterior region. For example, BE is the dimension
of the intersection between the boundary of object 1 and the exterior of object 2.

For Simple Features, GeoSPARQL specifies the topological functions for *equals,
disjoint, intersects, touches, crosses, within, contains*, and *overlaps*. All of them are
binary operators that take two geometry literal parameters and return Boolean values. As aforementioned, the first parameter's spatial coordinate system is used for
calculation. For Egenhofer relation, GeoSPARQL specifies functions for *equals,
disjoint, meet, overlap, covers, coveredBy, inside*, and *contains*. Some functions
such as the *intersects* function of Simple Features and the *meet* function of Egenhofer return true for multiple DE-9IM patterns.

For example, the *intersects* function of Simple Features relation, which is shown
below,

$$geof{:}sfIntersects(geom1{:}\ ogc{:}geomLiteral,$$

$$geom2{:}\ ogc{:}geomLiteral){:}\ xsd{:}boolean$$

corresponds to DE-9IM patterns

T********

*T*******

T**

****T****

These patterns indicate that two objects intersect, if the interiors of both objects intersect, or the interior of object 1 intersects the boundary of object 2, or the boundary of object 1 intersects the interior of object 2, or the boundaries of both objects intersect.

The RCC8 functions specified by GeoSPARQL represent more basic topological relations, which include *disconnected (DC), externally connected (EC), equal (EQ), partially overlapped (PO), tangential proper part (TPP), tangential proper part inverse (TPPi), non-tangential proper part (NTPP)*, and *non-tangential proper part inverse (NTPPi)*. Each of the RCC8 functions uniquely corresponds to a DE-9IM intersection pattern without wildcard.

Example 4.4 The following query retrieves the features that are contained in the city of Milwaukee using the extension function *geof:sfContains*. The prefix declarations are omitted from the example.

Select *?p*

Where *{*

?p	*geo:hasGeometry*	*?g* .
?g	*geo:asWKT*	*?w1*
gn:Milwaukee	*geo:hasGeometry*	*?m* .
?m	*geo:asWKT*	*?w2*

Filter *(geof:sfContains(?w2, ?w1))*

}

4.3.4 Topological Relations and Query Rewriting

In addition to the extension functions applicable to geometry literals in filter expressions, GeoSPARQL also specifies the predicates for directly asserting topological relations between spatial objects. There are three sets of predicates for the relation family of Simple Features, Egenhofer, and RCC8, respectively. These predicates directly correspond to the extension functions for topological relations in the previous section. To calculate the topological relation between two spatial objects, Geo-SPARQL specifies a set of query rewriting functions to translate the triple patterns involving topological predicates to the corresponding extension functions. The query rewriting functions are straightforward. Basically, a rewriting function translates a triple pattern with a topological predicate, such as

$$?f1 \quad geo{:}sfIntersects \quad ?f2$$

to an external call to the corresponding extension function:

$$\textbf{\textit{External }} (geof{:}sfIntersects \ (?g1, \ ?g2))$$

where $?g1$ and $?g2$ are the serializations of $?f1$ and $?f2$ (if they are geometries) or the serializations of the default geometries of $?f1$ and $?f2$ (if they are features), respectively.

Example 4.5 The query in Example 4.4 can be restated as follows using the topological predicate *geo:contains*.

> **Select** *?p*
> **Where** {
>
> | | ?p | geo:hasGeometry | ?g . |
> | | gn:Milwaukee | geo:hasGeometry | ?m . |
> | | ?m | geo:contains | ?g |
>
> }

When calculating the RDFS entailment relations for query answering, GeoSPARQL uses the query rewriting mechanism to translate triples with topological predicates to calls to extension functions for topological operations.

GeoSPARQL allows data to be properly indexed and queried from spatial RDF stores. In addition, it is intended to be interoperable with both quantitative and qualitative spatial reasoning systems (Battle and Kolas 2012). With a quantitative spatial reasoning system, GeoSPARQL explicitly calculates the distances and topological relations among concrete geometries of features. With a qualitative geospatial reasoning system, GeoSPARQL allows topological inferences for features whose geometries are either unknown or cannot be made concrete (Grütter and Bauer-Messmer 2007).

4.4 Spatial Data Indexing

Querying spatial data from a knowledge base with a large number of spatial objects is very inefficient if the spatial objects are stored as ontology instances. Spatial indexing algorithms such as R-tree (Guttman 1984), Quadtree (Samet and Webber 1985), and KD-tree (Bentley 1975) are often used in traditional spatial databases to improve query performance. With spatial indexing, spatial queries can be answered more efficiently since the number of the searched spatial objects is greatly reduced.

In this section, we briefly discuss three important spatial indexing algorithms: R-tree, Quad-tree, and Kd-tree.

R-tree and its variations (Sellis et al. 1987; Beckmann et al. 1990) are the preferred algorithms for spatial indexing. R-tree performs well for all kinds of spatial objects including points, lines, and polygons. It is a height-balanced tree where each spatial object is inserted into the tree leaf using its minimum bounding rectangle (MBR) as a guide. Each node of a R-tree has a MBR that encloses the MBRs of its children. The number of the children of each node is maintained within a range [m, M], so that if there is an overflow of children at a node the node is then split into two, and if there is an underflow the two nodes are merged. If the insertion of a spatial object causes the root node to split, then the tree will grow by adding a new root.

R-tree is a dynamic tree whose shape depends on the order in which the spatial objects are inserted. The MBRs of the children of a node may overlap so that search operations may require descending into multiple branches of a node. The height of a R-tree is $O(\log_m^N)$, where N is the number of spatial objects. Since each search operation may explore multiple branches of a node, there is no guarantee of time complexity in the worst case, but in practice the average time complexity is log-time. Heuristics are used in deciding sub-tree insertion and node splitting. Improvement on the heuristics has resulted in variations of R-tree algorithms, such as R*-tree, R+-tree, and X-tree.

Figure 4.2 shows an example R-tree, where each tree node has 2 to 3 children. In addition, each tree node has a MBR that contains the children of the node. For example, node R6 contains its children R13 and R14 while the MBR of R2 contains both R6 and R7. R-tree supports efficient spatial query. Suppose we look for the objects that intersect the green circle 1 in Fig. 4.2. We first check each child of the root to determine which one's MBR intersects the circle. Since only the MBR of R2 intersects Circle 1, we will recursively test the children of R2 and discover that the MBR of R7 intersects the circle. After examining the children of R7, we find that only R15's MBR intersects with the circle. Since R15 is a leaf, we return it as the result. In this search, we only check one child of each node as we descend the tree. However, this is not the general case. Suppose we look for the objects that intersect

Fig. 4.2 An example of
a R-tree (*bottom*) and the
spatial objects (*solid red
rectangles at the top*) that it
stores. The dashed rectangles
are the MBRs of the tree
nodes

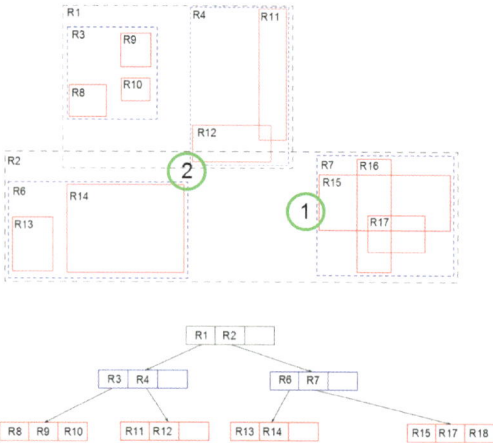

the green circle 2 in Fig. 4.2. Since Circle 2 intersects the MBRs of both R1 and
R2, we need to descend the sub-trees rooted at R1 and R2 and return the leaves R12
and R14.

Quad-tree is a tree data structure where each tree node has four children with
each child representing a quadrant in a two dimensional space. Spatial index for
points can be constructed using quad-tree by recursively adding tree nodes until
each leaf contains at most one point. Quad-tree-based index may not be balanced
depending on the distribution of spatial objects. The height of a quad-tree depends
on the smallest distance between two points. The time complexity of searching the
nearest neighbor of a point using a quad-tree is linear to the depth of the tree.

A common type of quad-trees is region quad-tree, which divides the space into
four equally-sized quadrants. Region quad-trees are suitable for indexing points.
There are other types of quad-trees (with more complexity) for indexing lines and
polygons.

Figure 4.3 shows an example of a region quad-tree that indexes points. In the
example, the points are not evenly distributed so that some quadrants of the quad-
tree have more leaves than others. In other words, the heights of the sub-trees are
not even, which can result in uneven performance for different searches. In extreme
cases, the depth of a quad-tree can be linear to the number of points. The advantage
of quad-tree, however, is its simplicity, and it can be efficient for evenly distributed
points. To build a quad-tree, we recursively divide a region into four quadrants until
each leaf quadrant contains no more points than the maximum. To search a quad-
tree for the points within a range, we recursively descend the tree nodes and only
visit quadrants that intersect the range. For each leaf quadrant we visit, we return the
points of the leaf that are within the range. For example, consider the green circle
in Fig. 4.3, the quadrants that we need to visit are marked by rectangles with round
corners and dashed lines. Only two points will actually be tested to see whether they
are in the circle and one of them is returned as the result.

Fig. 4.3 A quad-tree example for storing points, where each leaf contains at most one point

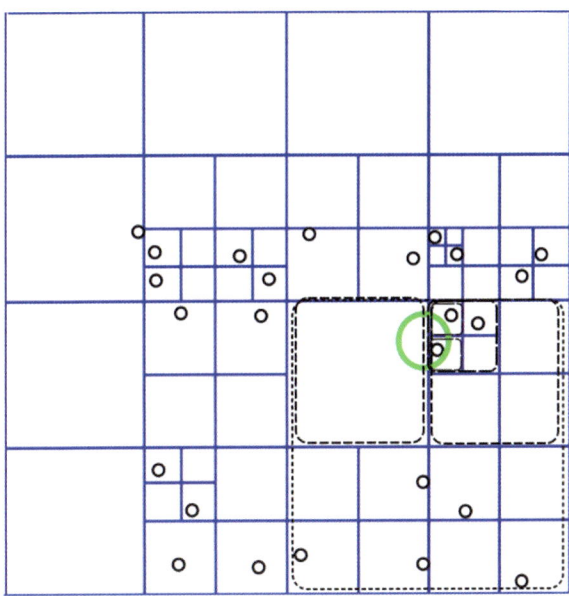

Kd-tree is a binary tree with a k-dimensional point in each node. For a node at depth d, the dimension $d \bmod k$ of the node's point is used as the key to separate the node's sub-trees. A value at each dimension (such as the medium value of the data points) needs to be chosen for data separation. Finding one nearest neighbor in a balanced kd-tree of randomly distributed points takes $O(\log n)$ time on average. However, in general, there is no guaranty that only one sub-tree of each node is explored in a nearest neighbor search.

In Fig. 4.4, we show an example of a kd-tree that stores points. The top of Fig. 4.4 shows the two dimensional layout of the points (labeled with letter a—h) and the splitting hyperplane (labeled with L1—L6) that divides the x or y space into two parts. The bottom figure shows the actual tree representation, where the intermediate nodes are the splitting hyperplanes that alternate between x and y dimensions and the leaves are the points. To construct the kd-tree for the points with label a—h, we first use the x dimension and separate a—d from e—h with the hyperplane L1. Then the y-dimension hyperplane L2 separates a, b from c, d while L3 separates e, f from g, h. The hyperplanes at the next depth level, L4—L7 are for the x-dimension again.

To search a kd-tree for points within a range, we use the splitting hyperplane to decide which sub-trees to visit. For example, in Fig. 4.4, the green circle is on the right side of L1 so that we visit the right side of L1, then we visit the bottom side of L3, and finally the left side of L6, which is a leaf node with point e. Since the point e is within the circle, it is returned as the result.

In summary, all three indexing methods can greatly reduce the runtime costs for range queries and nearest neighborhood queries of geospatial objects. However, quad-tree and kd-tree are more suitable for indexing point objects while R-tree and its variations are more convenient for indexing all types of spatial objects including lines and polygons.

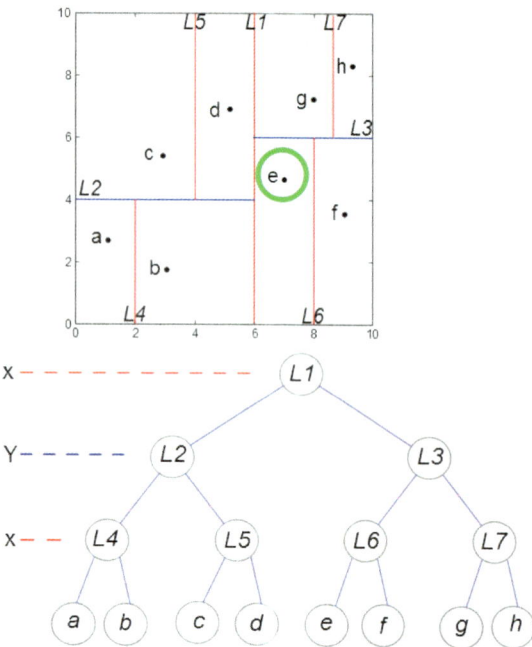

Fig. 4.4 A kd-tree example for storing points, where x and y dimensions are used to separate the points in the tree

4.5   Spatial Join

Spatial queries may require joining two or more types of spatial objects. In spatial databases, spatial join algorithms are used to improve runtime performance. The choice of algorithms depends on whether one or more spatial indices are present.

Nested loop (Mishra and Eich 1992) is the simplest approach to join two sets of spatial objects that may or may not have spatial indices. For each spatial object s_1 in the first set R_1, its spatial join relation with each spatial object s_2 in the second set R_2 is checked to determine whether the pair should be included in the result. If one of the two sets of objects is indexed, then it should be placed in the inner loop so that the index can be probed to find the object in the second set that satisfies the spatial join relation. The advantage of this approach is that it is very simple, does not require indices, and is efficient for small sets of spatial objects. This algorithm is easily parallelizable. The disadvantage is that it can be very slow for large data sets where the spatial relation is costly to compute. The algorithm is described by the following pseudo code.

> *for each* s_1 *in* R_1
> > *for each* s_2 *in* R_2
> > > *if* s_1 *is spatially related to* s_2
> > > > *output* (s_1, s_2)

Hierarchical traversal algorithms (Brinkhoff et al. 1993) can be used if both sets of objects are indexed using R-tree or similar data structures. The basic algorithm takes two sets of tree nodes *T1* and *T2* as inputs. For each pair of nodes with one from each set, the algorithm compares the MBRs of the nodes to determine whether they are disjoint. If they are not, then the algorithm recursively calls itself with the children of the non-leaf nodes as inputs. The recursion stops when both nodes are leaves and returns the pair of leaves that are related by the spatial join predicate. This approach is very efficient when the MBRs of a R-tree do not have many overlaps. Like nest loop, this algorithm is parallelizable. The drawback is the requirement of available indices on both sets of input objects, which may not be possible when the input sets are dynamically generated. The algorithm is described by the following pseudo code function *join*.

> *function* join *(T1, T2) {*
>
> > *for each* node n1 *in T1*
> >
> > *for each* node n2 *in T2*
> >
> > > *if* the MBRs of n1 and n2 are not disjoint
> > >
> > > > *if* n1 and n2 are leaves
> > > >
> > > > > *if* n1 and n2 are spatially related
> > > > >
> > > > > > *output(n1, n2)*
> > > >
> > > > *else if* n1 is a tree node, *call* join(n1.children, {n2})
> > > >
> > > > *else if* n2 is a tree node, *call* join({n1}, n2.children)

Plane sweep algorithm (Arge et al. 1998) does not use spatial indices. The algorithm works by sweeping along one dimension, e.g., x-dimension, so that if a MBR of a set *R1* is detected, it becomes active and is inserted into a sweep structure *S1* for that set. Once the sweep line passes a MBR, it is removed from the corresponding sweep structure. Each MBR in one sweep structure *S1* is compared with the active MBR of the other set *R2* to find out whether they are spatially related using other dimensions (e.g. y-dimension). Plane sweep algorithm works for smaller data sets that can be fitted into memory. The advantage of the approach is that it does not require either data sets to be indexed and it has better scalability than the nested loop algorithm when the objects are evenly distributed. The drawback is that it requires all the objects to be stored in memory for efficient processing and its applicability is limited by the size of the memory and the sizes of the data sets. The algorithm is described by the following pseudo code.

S1 = {}
S2 = {}
x = 0
a = min width of MBR

while *(x < max)*

      for each *r* **in** *R1, if r.MBR intersects x*

                 add r to S1

                 call *join(r, S2)*

      for each *r* **in** *S1, if r.MRB does not intersect x, remove r from S1*

      for each *r* **in** *R2, if r.MBR intersects x*

                 add r to S2

                 call *join (r, S1)*

      for each *r* **in** *S2, if r.MRB does not intersect x, remove r from S2*

      x = x + a

function *join(r1, S2) {*

      for each *r2* **in** *S2*

                 if *r1 and r2 are spatially related,* **output** *(r1, r2)*

}

For larger data sets, the partition-based spatial merge join algorithm can be used. This algorithm recursively partitions the pair of input data sets into pairs of smaller sets until each pair can be fitted in memory that the plane-sweep algorithm can be applied. The partition can be based on grid but MBRs that intersect grid lines must be duplicated.

To improve performance, the spatial join process can be split into two steps: filtering and refinement. The filtering step approximates the spatial join using the MBRs of the spatial objects to obtain a set of candidate pairs and the refinement step checks spatial relations using the actual geometries of the candidate pairs to obtain the final result. The filtering step processes more pairs of spatial objects but it is efficient to compute join relations for MBRs. The refinement step spends more time on computing the join relations of candidate pairs using their exact geometries though it involves a smaller number of spatial object pairs. The algorithms described above can be used for the filtering step.

4.6   GeoSPARQL Query Algorithm

While spatial indexing and spatial join algorithms are essential for efficiently processing spatial queries, the GeoSPARQL protocol does not specify whether and how they are adopted in its implementations. To consider how to incorporate spatial indexing and spatial joins in GeoSPARQL implementations, we consider an algorithm for answering GeoSPARQL queries defined with a simplified form of SPARQL syntax shown in Fig. 4.5.

A query Q consists of a set of selected variables v and a triple pattern P, which is a conjunction of a set of triple statements. A triple statement consists of a subject s, a predicate p, and an object o. The subject is either a variable or a URI, and a predicate is a URI, while an object can be a URI, a variable, or a literal. A URI includes a prefix and a short-name to identify an ontology resource. In this section, we only consider the URI that refers to an ontology class or a property. The answer to the query is a set of substitutions (Fig. 4.6), where a substitution σ maps each variable to a literal or a URI.

Fig. 4.5 Simplified syntax of GeoSPARQL

$$
\begin{aligned}
Q \quad &::= \quad \text{select } v \text{ where } P \quad &\text{query} \\
v \quad &::= \quad ?x \mid ?x \ v \quad &\text{selected variables} \\
P \quad &::= \quad s\, p\, o \quad &\text{triple patterns} \\
\quad &\mid \quad P_1\, P_2 \quad &\text{conjunction of triples}
\end{aligned}
$$

$$
\begin{aligned}
?x \quad &\in \quad V \quad &\text{variable} \\
s \quad &\in \quad U \cup V \quad &\text{subject} \\
p \quad &\in \quad U \quad &\text{predicate} \\
o \quad &\in \quad U \cup V \cup L \quad &\text{object}
\end{aligned}
$$

$$
\begin{aligned}
U \quad &: \quad \text{the set of URIs} \\
V \quad &: \quad \text{the set of variables} \\
L \quad &: \quad \text{the set of literals}
\end{aligned}
$$

Fig. 4.6 Definition of query solution

$$
\sigma : V \rightarrow U \cup L
$$

$$
\begin{aligned}
\sigma(s\, p\, o) \quad &= \quad \sigma(s)\, p\, \sigma(o) \\
\sigma(?x) \quad &= \quad l \quad &\text{if} \quad \{?x \mapsto l\} \subseteq \sigma \\
\sigma(l) \quad &= \quad l \quad &\text{if} \quad l \in U \cup L
\end{aligned}
$$

The query to a triple pattern P returns a set of σ (Eq. 4.1) such that the knowledge base (denoted by K below) entails $\sigma(P)$. We use the term of 'knowledge base' to refer the RDF data set that we use for answering GeoSPARQL queries.

$$query(s\ p\ o) = \left\{ \sigma | K\ \sigma(s\ p\ o) \right\}$$

where $K\ s\ p\ o$ means that the ontology K entails the triple $s\ p\ o$

$$(4.1)$$

The query of two triple patterns $P.\ P'$ returns the natural join of the answers to P and P' (Eq. 4.2).

$$query(P.\ P') = \left\{ \sigma \cup \sigma' | \sigma \in query(P) \wedge \sigma' \in query(P') \wedge F(\sigma, \sigma') \right\} \quad (4.2)$$

In Eq. 4.2, the function F ensures that the two solutions to a query are consistent.

$$F(\sigma, \sigma') = true \text{ iff } \forall\ ?x : \left(\sigma(?x) = l \wedge \sigma(?x) = l' \right) \Rightarrow l = l'$$

It may be very inefficient to process triple patterns P and P' separately since the answers to P and P' will be based on the entire knowledge base. However, if we process P first, we may be able to greatly reduce the solution space for P'. Therefore, it is more efficient to separate the triple patterns of a query into two sets with one set P_s for spatial queries and one set P_n for non-spatial queries. The separation is based on the predicates (spatial p_s versus non-spatial p_n) of the triples. The solution to P_n will be found first since it does not involve expensive spatial computations. Note that this strategy is used since it is often the case that the spatial query components are more computation intensive.

Thus, the query solution is redefined in Eq. 4.3 to use the solution $query(P_n)$ to non-spatial query P_n to restrict the set of solutions to each spatial triple $s\ p_s\ o$, and then join the results of spatial triples as the final solution. The solution for a spatial triple is defined by $K \vDash_s s\ p_s\ o$.

$$query(P_n.s\ p_s\ o) = \left\{ \sigma | \left(K_s \sigma \left(\sigma'(s\ p_s\ o) \right) \right) \wedge \sigma' \in query(P_n) \right\} \quad (4.3)$$

A knowledge base K entails a spatial triple $s\ p_s\ o$ (written as $K \vDash_s s\ p_s\ o$) if K entails a triple with subject s, a triple with object o, and s and o satisfy the relation $f_{p_s}(s, o)$, where f_{p_s} is a relation or function that implements p_s (Eq. 4.4).

$$K \vDash_s s\ p_s\ o \text{ iff } (K \vDash s\ \_\ \_) \wedge (K \vDash \_\ \_\ o) \wedge f_{p_s}(s, o) \quad (4.4)$$

For queries with multiple spatial triples, the results of each spatial query are joined in Eq. 4.5.

$$query(P_n.P_s^1.P_s^2) = query\left((P_n.P_s^1).(P_n.P_s^2) \right) \quad (4.5)$$

To find the solution for a spatial triple $s\ p_s\ o$, a strengthened entailment relation $\vDash_s$ is needed for the knowledge base. In particular, a query engine should implement spatial extensions f_{p_s} to perform the computations specified by the spatial predicate p_s. For example, to answer the query of the triple $?\, x\ geo : touches\ ?\, y$, it is necessary to know all of the spatial objects in the knowledge base that touch each other. To accomplish this, triple patterns with $geo : touches$ predicate are rewritten to the external function calls to the extension function $geof{:}sfTouches$. The extension function is written as $f_{geo:touches}$, which takes two parameters and returns true if the two parameters are spatial objects that touch each other. In a query engine implementation such as Jena library (https://jena.apache.org/), the extension functions such as $f_{geo:touches}$ are called for every possible solution to s and o. This is very inefficient.

The inefficiency comes from several sources. One is due to the fact that the geometries of spatial objects are stored as WKT or GML literals so that each time the objects are passed to the extension functions, the geometry literals have to be parsed. This is very costly for spatial joins. For example, if we want to find out pairs of objects that touch each other and there are N objects, then we have to make $\dfrac{N \times (N-1)}{2}$ calls to $f_{geo:touches}$ and the same N geometry literals are parsed $\dfrac{N \times (N-1)}{2}$ times. The extension functions are stateless and unable to cache the literals that have been parsed. Thus, to avoid this source of inefficiency, it is necessary to avoid passing geometry literals to the extension functions altogether. For example, the parsed values of the geometry literals can be cached and are used as inputs to the spatial functions that accept the parsed values.

Another source of inefficiency is due to the way a query is processed. In calculating the entailment of a knowledge base, the query engine will conduct an exhaustive search to find the triples that are answers to a triple pattern. A spatial index is not a consideration in GeoSPARQL specification. However, to answer a Geo-SPARQL query, it is possible to have pre-computed spatial indices for certain spatial objects so that the extension functions can take advantage of the indices to avoid a linear search.

For example, if we have indexed the geo-names in the knowledge base using an R-tree, then a query

$$?\, x\ geo : touches\ (43, -88)$$

can be answered by searching the index much more efficiently. However, this pre-computed spatial index approach only works in limited cases. For example, it can be used when the object of the triple query is a literal. For a query such as $?\, x\ geo : touches\ ?\, y$, the index may not be useful if the knowledge base contains multiple types of spatial objects. The reason is that the definition of *touches* relation depends on the geometries that are being compared while the available indices may be suitable for locating objects touching a point rather than a line or a polygon.

To allow more efficient query and yet remain flexible, it is possible to create on-the-fly indices for spatial objects. The actual indexing will be dependent on the spatial predicates and the geometry types of the spatial objects. For instance, if the goal is to find out the nearby *streets* of several *high schools*, we can index the *streets* based on x-coordinates so that *streets* are indexed to the closest *high school* based on the x-coordinates. After the indexing, the number of calls to $f_{geo:nearby}$ can be greatly reduced. Since the indexing is on-the-fly and specific to each spatial predicate, it can be implemented as a pre-processing function associated with $f_{geo:nearby}$. The knowledge base itself does not need to include any spatial indices.

The last source of inefficiency is the lack of parallelism in processing Geo-SPARQL queries. Even though triple statements can be processed in parallel with the results joined, the performance gain is not significant since the runtime of the triple queries is often uneven and in some cases, answering all triple statements sequentially may be faster than answering each triple statement in parallel if the triple statements are highly correlated. Thus, it may be more efficient to answer the non-spatial queries first and use their solutions to trim the solution space for the spatial queries.

4.7 Optimization and Parallelization

The implementation of an ontology query engine to answer GeoSPARQL queries can be improved by using the following optimization and parallelization strategies:

1. separating the non-spatial queries from the spatial queries,
2. caching the inputs to spatial extension functions to avoid repeatedly parsing geometry literals,
3. creating indices for certain inputs of spatial extensions on-the-fly, and
4. parallelizing the spatial joins using data partitioned based on the spatial indices, where a spatial join is represented by a triple of the form $s\ p_s\ o$ in the query and p_s is a predicate corresponding to a topological relation.

We focus on performance improvement for each spatial triple. Given a triple $s\ p_s\ o$, if either s or o is a literal, the query can be answered efficiently without parallelization. In the case where both s and o are variables, we first create indices for the spatial objects that are in any potential solution σ to s and o, then partition the solutions into k portions $\sigma_1, \sigma_2, \ldots, \sigma_k$, and finally compute $f_{p_s}(\sigma_i(s), \sigma_i(o))$ in parallel based on the partitions. The results of the parallel threads are then aggregated to form the final solution to $s\ p_s\ o$. For multiple spatial triple statements, we can execute them in parallel as well and join the results of the individual queries later. However, performance gain will be query dependent and it is not as predictable as parallelizing the query of an individual triple statement.

4.7.1 An Optimization Example

As a concrete example of optimizing the process of answering a GeoSPARQL query, consider a query $Q1$ that selects the nearby streets of each school in the state of Connecticut, where the schools are point features and the streets are poly-line features. As shown below, to decide whether there is a "nearby" relation between a point and a poly-line, the triple $?street\ geo:nearby\ ?school$ is rewritten as $?school\ ct:geom\ ?g_1, ?street\ ct:geom\ ?g_2$, and filter$(distance(?g_1, ?g_2) < 200)$. An extension function is defined to compute the distances between the spatial objects.

$Q1$: select $?school\ ?street$ where

    $?school\ rdf:type\ ct:school.$

    $?street\ rdf:type\ ct:street.$

    $?school\ geo:nearby\ ?street$

$\overset{rewrite}{\Rightarrow}$

select $?school\ ?street$ where

    $?school\quad rdf:type\quad ct:school.$

    $?street\quad rdf:type\quad ct:street.$

    $?school\quad geom\quad\quad ?g_1.$

    $?street\quad geom\quad\quad ?g_2.$

    filter$(distance(?g_1, ?g_2) < 200)$

The query processing workflow first rewrites the original query to expand certain triples to more primitive triples and filter statements using the pre-defined query rewriting rules. The primitive queries are grouped as spatial and non-spatial sub-queries. The solutions to the non-spatial sub-queries are indexed and partitioned for answering the spatial sub-queries in parallel. The query rewriting rules are pre-defined logic inference rules with each rule consisting of a head and a set of conditions.

For example, the rewriting rule for the 'nearby' relation is defined as follows.

$$?x\ geo:nearby\ ?y\quad :-\quad ?x\ geom\ ?g_1$$
$$?y\ geom\ ?g_2$$
$$\text{filter}(distance(?g_1, ?g_2) \leq 200)$$

This rule can be applied to $?school\ geo:nearby\ ?street$ since it can be unified with the head of the rule $?x\ geo:nearby\ ?y$ with the most general unifier $\sigma = \{?x \mapsto ?school, ?y \mapsto ?street\}$. The substitution σ is applied to the conditions of the rules and the results are added to the query. Note that definition of the *nearby* relation is subjective and the interpretation of the relation is manifested in the filter function of the rewriting rule, which is subject to a user's modification.

To parallelize the query, the query triples are grouped into two sub-queries: the first sub-query consists of the triples such as "$?school\ rdf:type\ ct:school$" and the second sub-query consists of the filter. In this example, the first sub-query is executed in parallel by sending triples of the same subject variable to the same thread.

The final results of the triple threads are aggregated as a set of solutions to the variables (*?school, ?g1, ?street, ?g2*). The results from the triple sub-queries are used for processing the filter sub-query.

The runtime of the GeoSPARQL query may be dominated by the filter sub-query. Therefore, it is possible to reduce the execution time of a query by increasing the number of threads for executing the filter functions. The cost of the triple sub-queries can be a large portion of the total query time so that the triples should be answered in parallel as well. The performance gain of the parallel execution can be limited by the underlying parallel architecture and the overhead of threading. The filter sub-query can be executed in parallel by dividing the inputs into equal proportions and sending them to each thread. For this sub-query, the inputs to p are divided into equally-sized blocks, where each block contains a N/p number of streets and N is the total number of streets and p is the number of threads. While the number of threads to execute a triple sub-query is bounded by the number of the subject variables involved, the number of threads to execute the filter sub-query is not bounded.

The cost of parsing the results of the triple sub-queries may be significant so that the results should be cached. In the RDF model, the geometries of the schools and the streets are represented as string literals of the forms $Point(x, y)$ and $LineString(x_1 y_1, x_2 y_2, \ldots, x_n y_n)$. The extension function for calculating object distances needs to parse the geometry literals into coordinates. In the optimized implementation, the literals can be parsed only once, and then be cached in object arrays. For a naive query engine, the geometry literals should be parsed each time that the distance function is called. The overhead of parsing is a major reason for the poor performance of the naive implementation. An alternative to reduce the cost of parsing is to represent the geometries of streets and schools as RDF instances with their coordinates as typed literals. However, it is difficult to define extension functions to operate on these RDF instances with a lower runtime overhead.

When calculating the distances between every pair of schools and streets, not all pairs of schools and streets need to be compared since it is only necessary to check the schools and streets that are relatively close to each other. Thus, spatial indexing can be applied to the inputs of the filter sub-query so that the inputs could be divided in a way that results in a more efficient parallel execution. A simple grid-based indexing algorithm is sufficient for this purpose. The algorithm first divides the schools into several sets based on their x-coordinates. For each street s, the algorithm checks if the x-coordinate of s falls into the range of the x-coordinates of all schools in a set (the range is extended with a buffer on each end to include all possible nearby streets). If it does, then the distances between s and each school in the set are calculated. For example, the schools can be divided into 10 sets, where each set contains schools that are close to each other. This may roughly reduce the number of comparisons between schools and streets by a factor of 10. The runtime of the filter sub-query can be reduced with more parallel threads. The scalability is not linear since the overhead of dividing up the sets of schools and partitioning the streets can erase the performance gain as more threads are introduced.

After applying the spatial indexing to a filter sub-query, however, the total runtime of the GeoSPARQL query may be dominated by the triple sub-queries. The performance of triple sub-queries could not be improved further without partitioning the RDF model, which is the subject of additional research.

4.8 Related Works

Different parallel approaches have been widely used for improving the query performance for a long time in literature. However, past research on improving query performance using parallelization has been centered on relational databases (e.g. Boral et al. 1990; DeWitt et al. 1986; Kitsuregawa et al. 1983). Optimizing techniques for parallel relational databases do not specialize on the triple model of RDF and triple patterns of SPARQL queries for query engines based on the RDF and SPARQL-specific properties (Groppe and Groppe 2011). Although there are studies to query heterogeneous relational databases using SPARQL and parallel algorithms (e.g. Miao and Wang 2009; Castagna et al. 2009; Karjalainen and Kemp 2009), parallel relational databases have inherent limitations such as scalability. SPARQL query can be parallelized by treating each triple statement in the query as a parallel task and the results of all the triple statement sub-queries are joined together after all the parallel tasks have been completed (Groppe and Groppe 2011). Unfortunately, this approach does not work efficiently when spatial predicates exist in the triple statements. There are also studies to propose methods for efficiently parallelizing joint queries of RDF data using Map-Reduce systems (e.g. Ravindra et al. 2011; Kim et al. 2011; Anyanwu 2013). However, to the best of our knowledge, there is no study to deal with parallelizing spatial join computations to support efficient spatial RDF query, which is an important issue for the development of a Geospatial Semantic Web (Zhao et al. 2008; Zhang et al. 2007, 2010a, 2010b, 2010c, 2013).

Query answering with large spatial data sets represented in Semantic Web languages consumes a large amount of physical memory and is computational intensive. As a result, distributed query processing strategies such as those using Map-Reduce frameworks to improve query performance become attractive options. Map-Reduce framework is a programming model for processing and generating large datasets (Dean and Ghemawat 2004). In the Map-Reduce model, programs written in the functional style are automatically parallelized and executed on a large cluster of commodity machines. The Map-Reduce concept was first proposed by Google to support its large distributed computing across a large number of machines on its huge databases (Dean and Ghemawat 2004). Recently, a number of studies have addressed the implementation of distributed SPARQL query engines using the Map-Reduce implementations such as the Hadoop framework (e.g. Husain et al. 2009; Choi et al. 2009; Kulkarni 2010).

Increasingly, distributed systems such as Cloud Computing (Cui et al. 2010; Liu et al. 2009, 2013; Yang et al. 2011a, b; 2013; Huang et al. 2013; Wen et al. 2013; Kim and Tsou 2013; Yue et al. 2013), cyber-GIS (Wang 2010; Wang et al. 2013) or spatial cyber-infrastructure (Wright and Wang 2011) have been suggested as a solution to overcome the scalability and performance problems of the Web-based GIS systems. Cloud Computing is a recent paradigm developed to search, access, and utilize large volumes of geospatial data for many geospatial science applications. Cloud Computing tools such as Hadoop and Spark are supported by many cloud service providers such as Amazon.

Hadoop implements the Map-Reduce model. Although Hadoop provides a simple and powerful mechanism to implement distributed applications while hiding details (e.g., instantiation of jobs in cluster, fault tolerance or data distribution), the binary relational operators (e.g., join, Cartesian product, and set operations) are difficult to implement in a pure Hadoop framework (Mazumdar 2011). Currently, Hadoop supports only partition parallelism in which a single operator executes on different partitions of data across the nodes. As a result, the existing Hadoop-based systems with the relational style join operators translate multi-join query plans into a linear execution plan with a sequence of multiple Map-Reduce cycles. This significantly increases the overall communication and I/O overhead involved in RDF graph processing on Map-Reduce platforms (Ravindra et al. 2011). In addition, files on Hadoop now cannot be modified randomly. This may limit many features such as update operation for RDF applications (Sun and Jin 2010).

4.9   Chapter Summary

Widespread use of geospatial applications demands flexible and efficient access to geospatial data. However, the traditional geospatial database is not sufficient to provide the mechanism for linking spatial data with non-spatial data over the web and to support flexible query. In this chapter, we provide a brief overview of the protocols SPARQL for ontology query and its extension GeoSPARQL, which enables geospatial applications to search for spatial features in ontology knowledge bases. GeoSPARQL queries are far more flexible than spatial database queries and WFS queries, but the flexibility comes with the cost of efficiency.

The geospatial knowledge base does not provide spatial indexing, which can result in a high computation cost for range queries, nearest neighborhood queries, and spatial joins. In this chapter, we explain three commonly used spatial indexing data structures: R-tree, quad-tree, and kd-tree. These indexing methods can greatly reduce the computation cost for range queries and nearest neighborhood queries, which is crucial to the usability of many geospatial applications. Geospatial queries may also involve spatial joins with a potentially high computation cost. In this chapter, we also introduce several algorithms for implementing spatial joins depending on whether spatial indices are available for the input spatial data sets. Again, geospatial ontology and GeoSPARQL do not provide efficient implementation for the spatial join operations.

To identify sources of performance bottlenecks in GeoSPARQL implementation, we explain a formal model for GeoSPARQL and describe how a query can be decomposed into several basic components so that they can be processed more efficiently. In particular, spatial joins can be processed more efficiently through caching, runtime indexing, and parallelization. Naive treatment of spatial objects in the ontology knowledge base results in repetitive parsing of ontology literal strings into spatial objects. By caching the parsed spatial objects, a GeoSPARQL implementation can greatly reduce the overhead in spatial join operations. Through runtime indexing, we can further reduce the cost of spatial join operations and also enable parallel execution of the spatial join computation.

References

Anyanwu K (2013) A vision for SPARQL multi-query optimization on MapReduce. In: IEEE 29th international conference on data engineering workshops (ICDEW). pp 25–26. doi:10.1109/ICDEW.2013.6547420

Arge L et al (1998) Scalable sweeping based spatial join. In: Gupta A et al (eds) Proceedings of the 24th international conference on very large data bases (VLDB). Morgan Kaufmann, New York, pp 570–581

Battle R, Kolas D (2011) Enabling the geospatial semantic Web with parliament and GeoSPARQL

Battle R, Kolas D (2012) Enabling the geospatial semantic web with parliament and GeoSPARQL. Semant Web J 3:355–370

Beckmann N, Kriegel H-P, Schneider R, Seeger B (1990) The R*-tree: an efficient and robust access method for points and rectangles. In: Proceedings of the 1990 ACM SIGMOD international conference on management of data (SIGMOD '90). ACM, New York, NY, USA, pp 322–331

Bentley JL (1975) Multidimensional binary search trees used for associative searching. Commun ACM 18:509–517

Boral H, Alexander W, Clay L, Copeland G, Danforth S, Franklin M, Hart B, Smith M, Valduriez P (1990) Prototyping Budda: a highly parallel database system. IEEE KDE

Brinkhoff T, Kriegel H-P, Schneider R (1993) Comparison of approximations of complex objects used for approximation-based query processing in spatial database systems. In: Proceedings of the 9th IEEE international conference on data engineering, Vienna, pp 40–49

Castagna P, Seaborne A, Dollin C (2009) A parallel processing framework for RDF design and issues. Technical report, HP Laboratories, 2009

Choi H, Son J, Cho Y, Sung MK, Chung YD (2009) SPIDER: a system for scalable, parallel/distributed evaluation of large-scale RDF data. In: Proceedings of the 18th ACM conference on information and knowledge management (CIKM 09), ACM, New York, NY, USA, pp 2087–2088

Cui D, Wu Y, Zhang Q (2010) Massive spatial data processing model based on cloud computing model. In: proceedings of the third international joint conference on computational sciences and optimization, Vol 2, pp 347–350, 28–31 May 2010, Huangshan

Dean J, Ghemawat S (2004) MapReduce: simplified data processing on large clusters. In: Proceedings of the 6th conference on Symposium on Operating Systems Design & Implementation (OSDI '04), Vol 6, USENIX Association, Berkeley, CA, USA, pp 137–149

DeWitt DJ, Gerber RH, Graefe G, Heytens ML, Kumar KB, Muralikrishna M (1986) GAMMA—a high performance dataflow database machine. In: Proceedings of the 12th international conference on very large data bases (VLDB '86), Wesley WC, Georges G, Setsuo O, Yahiko K (eds) Morgan Kaufmann Publishers Inc., San Francisco, CA, USA, pp 228–237

Groppe J, Groppe S (2011) Parallelizing join computations of SPARQL queries for large semantic web databases. In: Proceedings of the 2011 ACM symposium on applied computing (SAC '11). ACM, New York, NY, USA, pp 1681–1686

Grütter R, Bauer-Messmer B (2007) Combining OWL with RCC for spatioterminological reasoning on environmental data. In: Proceedings of third international workshop on OWL: Experiences and Directions (OWLED 2007), Innsbruck, Austria (6–7th June 2007) 258

Guttman A (1984) R-Trees: a dynamic index structure for spatial searching. In: Proceedings of the 1984 ACM SIGMOD international conference on management of (SIGMOD '84). ACM, New York, NY, USA, pp 47–57

Huang Q, Yang C, Benedict K et al (2013) Utilize cloud computing to support dust storm forecasting. Inter J Digital Earth 6:338–355

Husain MF et al (2009) Storage and retrieval of large RDF graph using hadoop and map-reduce. In: Jaatun MG et al (eds) CloudCom 2009, LNCS 5931. Springer, Heidelberg, pp 680–686

Karjalainen M, Kemp GJL (2009) Uniform query processing in a federation of RDFS and relational resources. In: Proceedings of the 2009 international database engineering & applications symposium (IDEAS '09). ACM, New York, NY, USA, pp 315–320

Kim IH, Tsou MH (2013) Enabling digital Earth simulation models using cloud computing or grid computing—two approaches supporting high-performance GIS simulation frameworks. Inter J Digital Earth 6:383–403

Kim H, Ravindra P, Anyanwu K (2011) From SPARQL to map-reduce: the journey using a nested triplegroup algebra. Proc VLDB Endow (PVLDB) 4:1426–1429

Kitsuregawa M, Tanaka H, Motooka T (1983) Application of hash to data base machine and its architecture. New Gener Comput 1(1):63–74

Kulkarni P (2010) Distributed SPARQL query engine using MapReduce. Master of Science Thesis, Computer Science School of Informatics University of Edinburgh, pp 18–31

Liu Y, Guo W, Jiang W et al (2009) Research of remote sensing service based on cloud computing mode. Appl Res Comp 26:3428–31

Liu Y, Sun AY, Nelson K et al (2013) Cloud computing for integrated stochastic groundwater uncertainty analysis. Inter J Digital Earth 6:313–337

Mazumdars (2011) Complex SPARQL query engine for Hadoop MapReduce. Master Thesis, Université de Nice Sophia Antipolis, 2011

Miao Z, Wang J (2009) Querying heterogeneous relational database using SPARQL. Eighth IEEE/ACIS International conference on computer and information science

Mishra P, Eich MH (1992) Join processing in relational databases. ACM Comput Surv 24:63–113

OGC 11-052r4 (2012) OGC Geo-SPARQL—a geographic query language for RDF data. http://www.opengis.net/doc/IS/geosparql/1.0. Accessed 17 May 2015

Ravindra P, Kim HS, Anyanwu K (2011) An intermediate algebra for optimizing RDF graph pattern matching on MapReduce. In: Proceedings of the 8th extended semantic web conference on the semanic web: research and applications (ESWC '11), Grigoris A, Marko G, Elena S, Bijan P, Dimitris P (eds) Vol. Part II. Springer-Verlag, Berlin, Heidelberg, pp 46–61

Samet H, Webber RE (1985) Storing a collection of polygons using quadtrees. ACM Trans Graph 4:182–222

Sellis TK, Roussopoulos N, Faloutsos C (1987) The R+-Tree: A dynamic index for multi-dimensional objects. In: Proceedings of the 13th International Conference on Very Large Data Bases (VLDB '87), Peter MS, William K, Peter H (eds) Morgan Kaufmann Publishers Inc., San Francisco, CA, USA, pp 507–518

SPARQL (2008) SPARQL query language for RDF. W3C recommendation, 15 Jan 2008

Sun J, Jin Q (2010) Scalable RDF store based on HBase and MapReduce. In: 3rd international conference on advanced computer theory and engineering (ICACTE), vol 1, pp 633–636, 20–22 Aug 2010

Wang S (2010) A cyberGIS framework for the synthesis of cyberinfrastructure, GIS and spatial analysis. Ann Assoc Am Geogr 100:535–557

Wang S, Anselin L, Badhuri B et al (2013) CyberGIS software: a synthetic review and integration roadmap. Inter J Geogr Info Sci 27:2122–2145

Wen Y, Chen M, Lu G et al (2013) Prototyping an open environment for sharing geographical analysis models on cloud computing platform. Inter J Digital Earth 6:356–382

Wright D, Wang S (2011) The emergence of spatial cyberinfrastructure. Proc Natl Acad Sci U S A 108:5488–5491

Yang C, Wu H, Huang Q et al (2011a) Using spatial principles to optimize distributed computing for enabling the physical science discoveries. Proc Natl Acad Sci U S A 108:5498–5503

Yang C, Goodchild M, Huang Q et al (2011b) Spatial cloud computing: how geospatial sciences could use and help to shape cloud computing. Inter J Digital Earth 4:305–329

Yang C, Xu Y, Nebert D (2013) Redefining the possibility of Digital Earth and geosciences with spatial cloud computing. Inter J Digital Earth 6:297–312

Yue P, Zhou H, Gong J et al (2013) Geoprocessing in cloud computing platforms—a comparative analysis. Inter J Digital Earth 6:404–425

Zhang C, Li W, Zhao T (2007) Geospatial data sharing based on geospatial semantic web technologies. J Spat Sci 52:11–25

Zhang C, Zhao T, Li W (2010a) Automatic search of geospatial features for disaster and emergency management. Inter J Appl Earth Obs Geoinfo 12:409–418

Zhang C, Zhao T, Li W et al (2010b) Towards logic-based geospatial feature discovery and integration using web feature service and geospatial semantic web. Inter J Geogr Info Sci 24:903–923

Zhang C, Zhao T, Li W (2010c) A framework for geospatial semantic web based spatial decision support system. Inter J Digital Earth 3:111–134

Zhang C, Zhao T, Li W (2013) Towards improving query performance of web feature services (WFS) for disaster response. ISPRS Inter J Geo-Info 2:67–81

Zhao T et al (2008) Ontology-based geospatial data query and integration. In: Geographic information science, lecture notes in computer science, vol 5266, pp 370–392

Chapter 5
Volunteered Geographic Information (VGI) systems and their interactions with Geospatial Semantic Web

5.1 Volunteered Geographic Information (VGI) and Geospatial Semantic Web

With the development of the Internet, especially Web 2.0, user-generated contents such as blogs, wikis, and social networks have gained a lot of attention and generated a lot of information over the Web. In geospatial community, the term "Volunteered Geographic Information (VGI)" was created to refer to any user-generated contents that have a relation to the surface of the earth (Goodchild 2007). VGI is the crowd-sourcing *geographic* data provided voluntarily by individuals. It is a special case of the larger Web phenomenon known as user-generated contents. Because the technologies used for VGI, such as GPS and the web 2.0 development including broad band communication, are mature, recently VGI has been used for a variety of applications and it is constantly growing now.

There are many benefits to use VGI for mapping or other applications. For example, VGI is an efficient means of acquiring timely and detailed geographic information. Many of other means of mapping are slow, especially for time-critical applications such as emergency response, location based services, and real time traffic management. The time-critical applications need instant access to spatial data to make quick decisions and take instantaneous actions. However, it is difficulty to acquire even basic geographic information promptly. Many existing spatial data and maps are outdated or would become outdated instantly if a special event happens. For example, a disaster can significantly affect the local geography, making the existing spatial data and maps outdated instantly. A highway may be closed because of an accident or a flood disaster. The dynamic nature of our world requires timely updating of a variety of data/information. However, it is challenging to update the data/information rapidly.

Although remotely sensed images can provide the timely information that is particularly useful, there are still some phenomena that cannot be sufficiently measured by both in-situ and remote sensing systems (Goodchild 2007). While other data sources such as remote sensing imagery provide much geographic information worldwide, many attributes of geographic features including place names cannot be seen from the remote sensing sensors. Other phenomena such as storms and the close-down status of hospitals cannot be measured by remote sensing systems, too. However, humans can obtain the information about phenomena that cannot be measured by remote sensing sensors or other methods. Human can obtain more detailed information about a phenomenon or an object, such as information about the inside of a building. For another example, human can identify air pollution or water pollution phenomena by using our special smell or taste organs.

In addition, measurements from sensors may not be available due to communication interruptions or the destruction of sensors (Goodchild 2007). For example, water gauges may be destroyed by very severe floods. Moreover, sensors may not be able to take measurements at serious moments. It is difficult to obtain needed information from optical remote sensing imagery if there are clouds in the imagery. Finally, a time delay may occur due to data acquisition and processing for remote sensing. Currently, it may take several days to get satellite imagery. This means that images cannot be acquired quickly when a time-critical event happens.

Some of these gaps may be filled by VGI. Comparing with intermittent coverage by professionals, VGI allows more timely observations by densely distributed amateurs for local changes. Volunteers can use VGI systems such as GoogleMaps and the Google API to create and disseminate their own maps in the real time over the Web. In addition, VGI is free while other mapping methods such as methods used to update the United States Geological Survey (USGS) maps are expensive. VGI can provide new data at a low cost. It's cheaper than other alternatives. Furthermore, products of VGI are almost invariably available to all Internet users. Internet technology has been widely used for sharing data/information because of advantages such as platform independence, reductions in distribution costs and maintenance, ease of use, and ubiquitous access and sharing of information by the worldwide user community. Compared with non-web-based approaches, the web-based VGI approach has advantages such as informing and engaging the public and encouraging the public to participate.

A significant progress has been made in developing the web-based VGI systems. OpenStreetMap (http://www.openstreetmap.org/) is one of the most well-known examples of VGI initiatives. The main goal of OpenStreetMap is to create a free editable map of the world. Through the OpenStreetMap interface on the Web, any user can upload new streets in the world through GPS tracks or modify existing information. So far OpenStreetMap has grown to over 1.6 million registered users, who can collect data using GPS devices, aerial photography, and other free sources. OpenStreetMap has provided a complex diversity of spatial data for the public use over the Web.

WikiMapia is another well-known example of VGI initiatives (www.wikimapia. org). WikiMapia combines an interactive web map with a geographically-referenced wiki system to allow users to mark and describe the geographical objects in the world. Through a map interface, the WikiMapia can allow users to create bounding boxes, or more detailed polygons inside a bounding box to provide information related to a geographic location such as towns, rivers, cities, and buildings. The users can insert a title, a short description, and a link to a Wikipedia page that gives more information about the geographic location. As of May 2014, over 23,000,000 objects have been marked by registered users and guests (Wikimapia.org 2014). The Wikimapia website provides a Google Maps API-based interactive web map that consists of user-generated information layered on top of Google Maps satellite imagery and other resources.

Google Maps (http://maps.google.com/) provides VGI information by allowing users to create personal maps through its geographic interface. Google Maps is a desktop and mobile web mapping service application provided by Google. Users can create point, line, and polygon maps and associated the maps with photos, videos, text, and html documents in Google Maps. Google Maps offers satellite imagery, street maps, and Street View perspectives, as well as functions such as a route planner for traveling by foot, car, bicycle or public transportation. Please notes that Google Maps satellite images are not updated in real time. However, Google adds data to their Primary Database on a regular basis (Anderson 2011). Other users can search such map contents via the Google search engine.

Flickr (http://flickr.com) also contributes tremendous amounts of place-based information via its members' shared photographs of specific locations. Flickr did not originally intend to provide geodata. However, because many photographs are related to specific locations, Flickr provides rich location-specific images as photographic information accumulates. The tremendous amounts of place-based images has undeniably improved geographic information and enhanced the overall body of environmental knowledge. Flickr can be seen as an important VGI for understanding a particular physical environment, and current information about local conditions provided by individuals, who in most cases are not trained or even not necessarily interested in geography as a science.

With GPS-enabled smart phones become popular, VGI information such as geotagged blog posts, messages, photos, or videos generated by the GPS-enabled smart phones becomes more ubiquitous. The Ushahidi project and the Sahana FOSS (Free and Open Source Software) Disaster Management System are two such smart phone based VGI examples. The Ushahidi project (http://www.ushahidi.com/product/ushahidi/) provides tools that can collect eyewitness reports by emails and text messages from smart phones and place them on a "Google Maps" map. Local observers of events, such as crisis, disaster, and violence, can submit their reports using their mobile phones to create a temporal and geospatial archive of the events. Ushahidi gives users the option of using OpenStreetMap maps in its user interface and requires the Google Maps API for geocoding. Its main products include SwiftRiver and Crowdmap. SwiftRiver is a suite of intelligence and real-time data gathering products that complement Ushahidi's mapping and visualization products. Crowdmap is a simple map-making tool built on an open API to allow users to collaboratively map the world over the Web.

The Sahana FOSS Disaster Management System is another implemented example of VGI systems, which was initially developed to help manage the disaster during the 2004 Sri Lanka tsunami. "Sahana" means "relief" in the Sinhala language. The system was initiated by considerable relief coordination needs in Sri Lanka following the tsunami. The project has grown since then, and it has been deployed during other disasters such as the Asian Quake in Pakistan in 2005, Southern Leyte Mudslide Disaster in Philippines in 2006, and the Jogjakarta Earthquake in Indonesia in 2006. With the Sahana system the disaster coordination and relief hub may be located away from the affected disaster region, and users can use satellite communications or mobile phones for the disaster coordination and relief purposes. Sahana includes a synchronization mechanism to allow individuals to go to the field with a snapshot of existing Sahana data and later synchronize with the central server when Internet connectivity is available.

In general, VGI has the potential to be a great source for us to understand the surface of the Earth. It has proven very useful in acquiring timely and detailed geographic information for many applications. Several rapid and successful VGI deployments employing collaborative web-based systems such as Open Street Map, Google Earth/Maps, Ushahidi, Sahana, and Flickr have been used to assist in disaster response situations (e.g. Laituri and Kodrich 2008). The web-based VGI software helped to coordinate disaster responses recently, such as the earthquake that struck Haiti in January 2010, Gulf of Mexico Oil Spill in April 2010, and the series of damaging wildfires that affected Santa Barbara between 2007 and 2009 (Goodchild and Glennon 2010). The launch of the Ushahidi platform in Haiti demonstrated the potential of using mobile technology for information gathering and communication during disaster response and management. Several case studies have shown the added-values of using VGI in various types of crisis events (De Longueville et al. 2011), such as earthquakes (De Rubeis et al. 2009), forest fires (De Longueville et al. 2009), political crises (Bahree 2008), hurricanes (Hughes and Palen 2009), floods (De Longueville et al. 2010), and terrorist attacks (Palen et al. 2009).

Although these VGI (or crowd-sourcing) applications have proven useful for gathering information about a crisis, these applications only provided limited utilities for response coordination (Gao et al. 2011). This is mainly caused by lack of compatibility within different software packages and heterogeneity among different data sources (Poser and Dransch 2010; Zook et al. 2010; Morrow et al. 2011). One major problem of VGI is the existence of the heterogeneous semantic problem. The challenges of semantic interoperability have been recognized in literature (e.g. Zhang et al. 2010a, b, c). Problems arise when VGI is to be used as decision-support information because the user-generated annotations are often unstructured and their contents are not machine-readable. It is difficult to use the user-generated data to support spatial analysis required in decision-making processes because computers do not understand the contents of the VGI. For example, with the currently deployed VGI systems, it is not easy to get answers to the questions:

> Which roads have been blocked (thus cannot be accessed) by the hurricane Irene disaster in the city of New Haven, Connecticut? And which is the best evacuation route?

The information required to find the answers cannot be readily retrieved because the information needs to be extracted from multiple data sources and the extraction process is time consuming. It may take several hours or days for a professional GIS person to collect the needed information, update the existing spatial database, and perform the spatial analysis to find out the best evacuation route.

Therefore, a semantic-empowered analysis tool, which is able to automatically integrate and analyze a broad variety of spatial data contents, is needed to provide more utilities for response coordination. This chapter intends to introduce such a VGI system based on Geospatial Semantic Web technologies for the public, who usually do not have many GIS skills, to update the existing databases and automatically search for the needed information. We present such an interoperable online VGI system based on state-of-the-art Geospatial Semantic Web technologies to overcome the limitations of the currently implemented VGI systems.

5.2 A Framework of an interoperable online VGI system

Figure 5.1 shows the framework of an interoperable online VGI system based on Geospatial Semantic Web technologies. The framework is able to overcome the semantic heterogeneity problems of the currently implemented VGI system and make use of the unstructured citizen-generated contents by using Geospatial Semantic Web technologies. Because VGI is produced by individuals who may use natural languages instead of formal languages used by existing geospatial databases and services, it may have higher levels of semantic heterogeneity than traditional official data. Although VGI may provide a lot of free data, it is difficult to integrate, share, or use these data without resolving the semantic heterogeneity problems, because it is impossible for the public to use consensus terminologies or standard

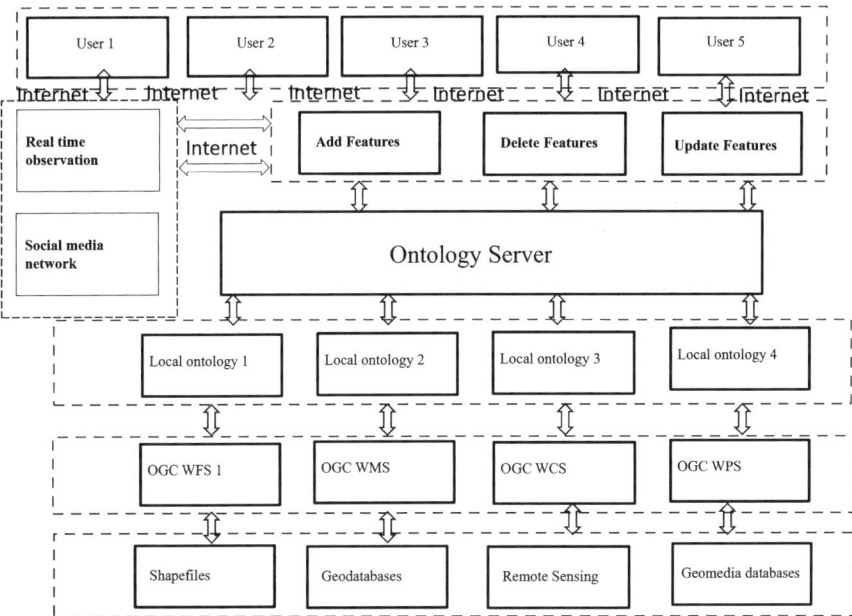

Fig. 5.1 The framework of an interoperable online VGI system

formats to produce VGI. The framework aims to facilitate integration, share, and use of VGI by resolving the semantic heterogeneity problems via Geospatial Semantic Web technologies.

A distributed local-responsible web service architecture is used in the proposed framework. OGC web services, such as Web Feature Services (WFSs), Web Map Services (WMSs), and Web Coverage Services (WCSs), are used to overcome the heterogeneous problem at the syntax level from different local geospatial databases, such as shapefiles and geodatabases in ArcGIS, remotely sensed satellite images in GeoTIFF format, and Geomedia databases. Ontology is used to add computer processable meaning (semantics) to the online VGI system over the World Wide Web. Ontology formally represents knowledge as a set of concepts and the relationships between those concepts within a domain and supports reasoning about concepts. Ontology has the mechanisms to resolve the semantic problem. The OGC web services are connected to local ontologies through local data source adapters. Heterogeneous local ontologies are integrated into an ontology server for web service discovery and integration. Through the proposed framework users such as the public, who usually have not much GIS knowledge and skills, can update the existing heterogeneous GIS databases from a variety of disparate sources through friendly graphic interfaces based on their real time observations or other knowledge, such as mining information from social media network (e.g., twitter, facebook, chat, skype, blogs, and emails).

Comparing with the existing implemented VGI systems, the major advantage of the framework is that it allows the public to share, integrate, and query the VGI at the semantic level. Systems developed based on this framework permit the public to remotely update (add, delete, change) semantically heterogeneous GIS databases, and automatically query and integrate spatial data from a variety of different sources based on data contents. In such an infrastructure each local ontology server offers lookup for local geospatial concepts within its spatial scope. The local ontology server should be maintained by the local community running the service. Thus, the stored ontology can be accurate and always the most updated. The main technologies applied in the framework are: (1) remotely updating spatial data, (2) matching geospatial features to the predefined geospatial ontologies, and (3) heterogeneous geospatial ontology integration. Since we have introduced algorithms for matching geospatial features to the predefined geospatial ontologies in Chapter Two, we focus on introducing (1) information about remote update of spatial data and (2) algorithms for heterogeneous geospatial ontology integration in this chapter.

5.2.1   Remotely updating spatial data

OGC Web Feature Services (WFSs) are used in the aforementioned framework to allow users to remotely update the outdated existing spatial data distributed in different sources in real time. As introduced before, WFSs manipulate spatial data at the feature level based on OGC's simple features (e.g. points, lines, and polygons) (Zhang et al. 2003; Peng and Zhang 2004; Zhang and Li 2005). Open-source standards such as XML (Extensible Markup Language) and GML (Geography Markup

Language) are used to represent features in WFSs. GML stores spatial data in text format, which is a vendor-neutral universal format. Because GML is not locked into a proprietary binary format, it is easy to integrate GML data into other data across a variety of platforms and devices. As introduced in Chapter One, the proprietary spatial data systems can be transferred into the standard GML in WFSs through *DataStores*. The *DataStores* can transform a proprietary data format such as ESRI's Shapefiles into the GML feature representation.

As introduced in Chapter One, to support remotely updating spatial data in the *DataStores*, five operations are defined in the OGC WFS: *GetCapabilities*, *DescribeFeatureType*, *GetFeature*, *Transaction*, and *LockFeature*. *GetCapabilities* describes the capabilities of the WFS server, such as which feature types it can serve and what operations are supported on each feature type; *DescribeFeatureType* informs the structure of any feature type upon a request; *GetFeature* retrieves feature instances; *LockFeature* processes a lock request on one or more instances of a feature type for the duration of a transaction; *Transaction* provides transaction requests for such operations on features as *create, update,* and *delete*. Users can manipulate geospatial data remotely at the feature level using the WFS capability of creating, deleting, and updating features. The feature level data manipulation can provide the most updated data for conducting further spatial analysis, modeling, and other operations. For example, the public can instantly edit a road feature to the *flood* status in their remote databases by using the *update* capability of WFS over the web, and they also can add a new *flood* affected location to their remote databases by using the *create* capability of WFS over the web.

Although the OGC WFSs provide capabilities of remotely updating geospatial data, the currently implemented OGC WFSs have limitations: They only emphasize technical data interoperability via standard interfaces and cannot resolve semantic heterogeneity problems in spatial data sharing (Zhang et al. 2010a). Thus, to achieve the goal of remotely updating spatial data at the semantic level over the Internet, in the proposed framework we extend the existing OGC WFS with geospatial semantic web technologies. We map OGC WFS descriptions to OWL ontologies to provide a semantically based view of the web services, which spans from abstract descriptions of the capabilities of the services to the actual feature data contents that exchange with other services. We focus on specifying semantic descriptions for three operations: *GetCapabilities*, *DescribeFeatureType*, and *GetFeature*. We specify semantics, meaning of languages (such as words, phrases), for each *FeatureType* and *FeatureProperty* in defined *GetCapabilities*, *DescribeFeatureType*, and *GetFeature* operations. The semantics of *FeatureTypes* and *FeatureProperties* in WFSs are mapped to disjunctions or conjunctions of (possibly negated) concepts in OWL ontologies. Because the definitions of the semantic concepts in the enhanced semantic WFSs are available at the referenced uniform resource identifier (URI) ontology database on the web, the WFS providers and the clients have a means of sharing terms. The results are that, by taking an OWL description of the WFSs, a WFS client can distinguish and properly interpret all *FeatureTypes* and *FeatureProperties* in *GetCapabilities*, *DescribeFeatureType*, and *GetFeature* operations in WFSs. VGI updated through the enhanced semantic WFSs allows the public to share, integrate, and query the VGI at the semantic level, thus it can overcome the semantic heterogeneity problem faced by the existing implemented VGI systems.

5.2.2   *Heterogeneous geospatial ontology integration*

In our prior work, we have used a single-domain ontology to ensure semantic in-teroperability (Zhang et al. 2007; Zhao et al. 2008). However, it is impractical to develop a global ontology for all disaster management applications that support the tasks envisaged by a distributed environment like the Geospatial Semantic Web. To overcome this problem, in the proposed framework of an interoperable online VGI system we adopt a distributed local-responsibility service infrastructure. Such a ser-vice infrastructure is an environment with multiple independent systems, of which each has its own local ontology. The advantage of this approach is that the local ontology can be designed to suit the geospatial data in each system. However, this approach brings the possibilities of conflicts and mismatches among different local ontologies. Thus, it is necessary for this study to develop algorithms to integrate the heterogeneous local ontology.

In IT literature, many schema alignment methodologies and software tools have been developed to integrate the heterogeneous ontologies (e.g. Shvaiko and Euzenat 2005; An et al. 2005). Although in IT literatures, algorithms and tools have been proposed to resolve the problem of heterogeneous ontology integration (e.g. Cas-tano et al. 2006), they are not developed for dealing with *spatial* data. Hess et al. (2006) recently proposed the G-Match algorithm for geographic ontology integra-tion. To match and integrate two different geographic ontologies, the G-algorithm measures overall similarity *Sim(C1, C2)* of their concepts by combining similarity measures of concept names, attributes, taxonomies, and conventional as well as topological relationships in a weighted sum. The G-Algorithm rejects a matching of two ontology classes if the overall similarity measure is below a predetermined threshold. One problem with this approach is that classes or properties with very similar names could be considered equivalent even though they are not in reality or they are not compatible in structure so that translation is impossible. Also, this ap-proach does not consider the range types of relations in computing their similarity.

In the proposed framework of an interoperable online VGI system, we adopt a Partition-Refinement algorithm (Zhang et al. 2010b) for integrating heterogeneous ontologies.

Definition 1. *An ontology consists of a set of classes and properties. A property can be a datatype property with the range of a primitive type or an object property that relates two class instances.*

Given two ontologies local ontology O_1 and sever ontology O_2, ontology mapping is defined as for each concept $C_1 = (T_1, A_1, R_1)$ in local ontology O1, finding a cor-responding concept $C_2 = (T_2, A_2, R_2)$, which has the same or similar semantics, in sever ontology O_2 where C_i is the class (or concept) in ontology O_i, Ti is the name of C_i, Ai is the set of datatype properties (or attributes) of C_i, R_i is the set of object properties (or relations) of C_i including spatial relations, and I = 1, 2.

We do not have to distinguish which ontology the classes and properties belong to when comparing two ontologies for similar classes and properties. Therefore, in the proposed partition-refinement algorithm, we first pool all classes and properties from the two ontologies together to form equivalence partition $P = \{p1, p2, ..., pn\}$, where p_i is a set of concepts or properties that are considered equivalent. Then we compute the equivalence relation based on similarity measures. Because a partition usually contains classes or properties that are similar but not exactly equivalent, we will adjust the similarity measure to refine the partition if necessary. Our similarity measure uses a structural equivalence as defined below.

Definition 2. *Ontology classes C1 and C2 are structurally equivalent if there is a one-to-one correspondence of structurally equivalent datatype properties between attr(C1) and attr(C2), and there is one-to-one correspondence of structurally equivalent object properties between rel(C1) and rel(C2), where attr(Ci) is the set of datatype properties of Ci and rel(Ci) is the set of object properties of Ci, for I = 1, 2.*

A datatype property a_1 is structurally equivalent to another datatype property a_2 if they have the same range type.

An object property r_1 is structurally equivalent to another object property r_2 if there is a one-to-one correspondence of structurally equivalent classes between range(r_1) and range(r_2), where range(r_i) is the set of range types of r_i, for i = 1, 2.

Since there are only a few primitive types, many datatype properties could have the same range types. Name similarity measure is applied to distinguish different datatype properties. Name similarity can be computed with the help of *WordNet* to find synonyms. Name similarity can be considered for object properties and classes as well, but in this case we need to find structurally equivalent object properties and classes first, and then apply the name similarity measure if too many equivalent classes or properties are found.

The name similarity measure is based on a reduced form of the string comparison measure (Stoilos et al. 2005). The similarity of two strings s_1 and s_2 is $Sim(s_1, s_2) = Comm(s_1, s_2) — Diff(s_1, s_2)$, where $Comm(s_1, s_2)$ measures the total length of the maximum common substrings of s_1 and s_2 divided by the total length of s_1 and s_2, and $Diff(s_1, s_2)$ is a function of the length of unmatched substrings of s_1 and s_2 scaled by the length of s_1 and s_2, respectively. $Comm(s_1, s_2)$ and $Diff(s_1, s_2)$ range between 0 and 1. Thus, the similarity value ranges from -1 to 1. The threshold of accepting similar pairs of names can be chosen between 0 and 1, but it needs to be adjusted based on actual data.

We also consider the spatial characteristics of ontology classes to find initial partitions. In particular, we infer the geometry type of an ontology class to determine whether it is *Point, LineString, Polygon, MultiPoint, MultiPolygon*, or *GeomCollection*. Given a geometry ontology class and its instances, it is possible to distinguish its geometry type based on its spatial contents. For example, a *Point* class has at most a pair of x, y coordinates while other geometries have multiple pairs of such coordinates. *LineString* and *MultiPoint* geometries are different from *Polygon* because the latter one must form a closed loop. *MultiPolygon* and *GeomCollection* are different from the rest because they must use delimiters to contain multiple geometries while *MultiPolygon* contains only one type of geometries and *GeomCollection* includes various kinds of geometries. Only *LineString* and *MultiPoint* cannot be easily distinguished since both contain a sequence of points. So they are assigned to the same initial partition and need to be separated using a partition refinement algorithm. Another kind of spatial characteristics that we consider are spatial relations, which include *within, overlap, contain, touch, intersect, disjoint, cross,* and *equal.* Though the spatial relations may be computed, sometimes they are specified in ontology as well to improve performance. In ontology, these relations are treated as object properties. To identify them, we use name comparison and also utilize spatial constraints to eliminate properties that cannot be spatially related. For example, if we let P be *point,* L be *line,* and A be *polygon,* then the *within* relation applies to *P/L, P/A, L/L, L/A,* and *A/A* groups of relationships and the *cross* relation applies to *P/L, P/A, L/L,* and *L/A* situations. Once we identified spatial relations, we replace the *within* relation with *within-*1 by switching its domain and range. The notation *within-*1 represents a property similar to *within* but with its domain and range reversed and it is considered equivalent to *contain.* Also, since *intersect* is more general than *overlap,* we allow them to be matched in ontology alignment.

The following is a basic version of the Partition-Refinement algorithm:

Input: *A set of classes, datatype properties, and object properties.*

Output: *Partitions of classes P_C, object properties P_R, and datatype properties P_A, where each partition in P_C contains equivalent classes and each partition in P_A (P_R) contains equivalent datatype (object) properties.*

Step 1: *initialization:*

    a. Divide the set of datatype properties into P_A where each partition in P_A contains datatype properties with similar names and the same range type.

    b. Put the set of object properties into P_R with one partition for spatial properties and one partition for the non-spatial properties. Note that at initialization, it is not necessary to separate spatial properties from the rest but this can help increase the precision of the algorithm.

    c. Divide the set of classes into P_C such that C_1, C_2 is in a partition in P_C if and only if for each partition p in P_A, $|p \cap attr(C_1)| = |p \cap attr(C_2)|$, where $|p \cap attr(C_i)|$ is the size of the set $p \cap attr(C_i)$.

Step 2: *partition refinement:*

    a. Refine P_R such that r_1, r_2 are in a partition in P_R, if and only if for each partition p in P_C, $|p \cap range(r_1)| = |p \cap range(r_2)|$.

    b. Refine P_C such that C_1, C_2 are in a partition in P_C, if and only if for each partition p in P_R, $|p \cap rel(C_1)| = |p \cap rel(C_2)|$.

Step 3: *repeat Step 2 until P_R and P_C stabilize.*

Step 4: *refine P_C and P_R even further based on the name similarity of the classes and object properties. If this step results in any changes, then go back to Step 2.*

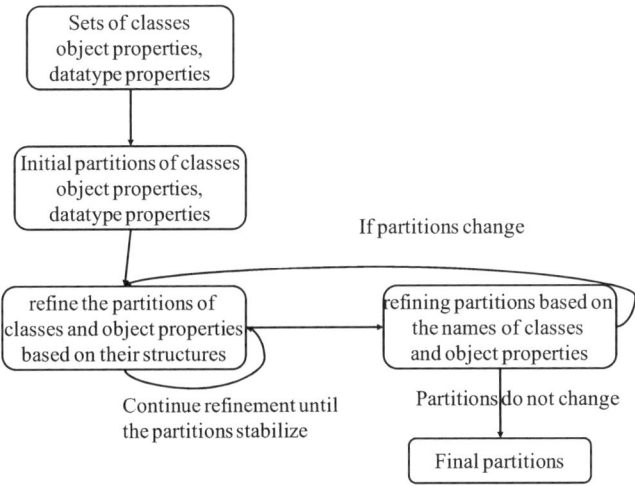

Fig. 5.2 The process of finding equivalent ontology classes and properties through partition refinement (adapted from Zhang et al. 2010b)

Figure 5.2 illustrates the process of finding equivalent ontology classes and properties according to the algorithm shown above. First, we pool all sets of classes, object properties, and datatype properties from both local and server ontologies. Then we initialize partitions of classes, object properties, and data type properties to form equivalence partitions. After that, we refine the partitions of classes and object properties based on their structures. If too many equivalent classes and object properties are found, we refine the partitions based on names of classes and object properties. The refinement process will be continued until the partitions become stable. Finally we obtain the final stable partitions, which will not change and are considered to be equivalent ontology classes and properties.

The above algorithm always terminates, because each iteration splits at least one set in the partitions of classes or object properties. This problem is similar to finding structurally equivalent recursive types for object-oriented programming languages and the time complexity was shown to be $N \, log \, N$, where N is the total number of classes and properties (Jha et al. 2002).

Note that the partition refinement step can be adjusted to allow a partition to include more (or less) similar classes or properties. For example, instead of requiring $|p \cap rel(C_1)| = |p \cap rel(C_2)|$ for the refinement of class partitions, we can require only $|p \cap rel(C_1)| \approx |p \cap rel(C_2)|$, where $\approx$ is 'approximately equal' so that two classes are structurally equivalent if they contain approximately the same number of equivalent object properties (e.g. within 1 or 2).

Comparing with the G-Algorithm proposed by Hess et al. (2006), the main advantage of the partition refinement algorithm is that it finds matching ontology classes and properties based on their structures. Unlike the G-Algorithm, which heavily relies on a string-based similarity measure, the partition refinement algorithm makes full use of the structures of the ontology being mapped. Thus, it allows translation of instances among different ontologies. Further, the partition refinement algorithm can deal with recursive structures efficiently while the G-Algorithm cannot. We know that ontology classes may form recursive data structures where, for example, a class A may have an object property with the range of another class B, which may include a property that points back to class A. While the partition-refinement algorithm can find matching classes with this recursive structure efficiently, G-Algorithm is unable to deal with recursively defined ontology classes. This is due to the fact that G-Algorithm uses a top-down approach to compute similarity scores between two classes (or concepts) using their names, datatype properties (attributes), and object properties (relations). Because G-Algorithm needs to compute the similarity for object properties first before deciding whether classes can be matched, it is unable to deal with recursively defined ontology classes. Another problem with the G-Algorithm approach is that the threshold values are hard to determine and so are the weights assigned to different components of the score. If the weight for name similarity is too high, it is likely that the matched ontology classes are semantically different but they just have similar names by accident, while classes with similar semantics are not matched due to their different names. Thus, the G-Algorithm fails to discover some correct mappings when the mapped ontologies are syntactically dissimilar. Our algorithm, however, can overcome this shortcoming and may discover meaningful and syntactically unidentifiable mappings. Note that our algorithm also considers names, especially for datatype properties, because their range types are very limited. However, the class and object property names are secondary to their structures. The names can be used when too many matches are found.

5.3   A Case Study for Developing an Interoperable Online VGI System

Connecticut experienced many significant storms in the recent past. For example the August Tropical Storm Irene hurricane and the October Nor'easter hurricane in 2011 caused large damages and left a record number of residents (almost 1 million residents) without electricity, heat or reliable supplies of water for up to 9–12 days. Irene downed approximately 1–2 % of the State's trees. Damage from both storms was estimated to be $ 750 million–$ 1 billion.

The significant impact of these storms has served as a wake-up call to Connecticut. After the storms, the State Governor created a Two Storm Panel to review the preparedness, response and recovery efforts during the Irene and Nor'easter storms. Based on the final report of the Two Storm Panel (http://www.governor. ct.gov/malloy/lib/malloy/two_storm_panel_final_report.pdf), two of the important reasons preventing fast responses to the two disasters are: (1) the inefficient collaboration between municipalities, state resources, and electric utilities service providers, and (2) the dissatisfied communication between labor and management in all utilities. One important reason that caused the disappointed collaboration and dissatisfied communication is the lack of sharing updated data/information. The outage maps provided during the two storms did not provide local details. Questions as to which streets were blocked, what poles and wires were down, where the power was on and where it's off were consistent complaints during the two storms. Although the utility companies provided information on line breaks based on their existing grid system, they relied on consumers to identify exactly where resulting problems existed. However, there is no system to allow consumers share the information with them. In addition, based on the final report, some text messaging services offered by utilities directed individuals to shelters that had been closed or moved. The text information was not updated in a timely fashion.

In this case study, we intend to develop a prototype based on the aforementioned framework to overcome some of the problems met during the two storms in Connecticut. We anticipate that the prototype will provide useful tools to allow users create and update data/information over the Internet, thus they can provide timely, accurate, and up-to-date information for better disaster response. The system should facilitate real-time sharing of the existing legacy data and the new updated information among different organizations such as town responders, utility companies, medical departments, and police departments, so that they can work together and coordinate seamlessly in the event of a disaster. The contents of the implemented prototype should also be machine-readable so that computers can understand the content of the databases; thus, the required information for disaster response can be readily retrieved from multiple data sources.

The prototype is accessible from the website: http://boyang.cs.uwm.edu:8080/ newHaven/ and its implementation details are coved in Chap. 6.

The prototype can not only dynamically update new sources of information but also integrate heterogeneous information at the semantic level. It, therefore, can be used by users including disaster responders and the local residents to report information about events and damages in their neighborhood and update the related spatial information over the Web. Users of the prototype need only to understand the definitions for general concepts of streets, roads, and places to be able to query and update the spatial information; they don't need to understand the underlying definitions of the spatial data, which may be different across regions or across data sources. As long as the definitions of the original spatial data are properly mapped to the ontology-based definitions, ontology-based queries can be understood by the prototype.

The prototype is capable of turning the people with smart phones or internet access into a big human sensor network for fast updating the existing databases for disaster response. Using a mobile client, a citizen can report disaster events, such as the failure of a bridge (for example, Q Bridge), by simply touching a smart phone screen. As shown in Fig. 5.3, the system can automatically highlight the affected streets and roads that cross a failed bridge.

In summary, the implemented prototype can allow users share updated information over the Web. It can allow users provide local detailed information about the disaster situation. For example, the information about which streets are blocked,

SELECTED PLACES

NAME CATEGORY   TOWN   TYPE

Q Bridge   Bridge   New Haven bridge nearby streets nearby roads

Fig. 5.3 The affected roads across a failed bridge

what poles and wires are down, where the power is on and where it's off, can be updated and shared using the implemented online interoperable VGI prototype.

5.4 Current Challenges and Future Studies for Online VGI System

VGI enables a vast number of anonymous contributors to create spatial data over the Web. It is open access. Anyone can add or edit spatial information via VGI with no proof of identity or qualification. VGI can facilitate updating the spatial databases within a short time period with no or little cost. It can be used to enhance the information used by professionals and decision makers, and can provide users with local, detailed and spatial information that can be used to increase opportunities for public participation in the decision and policy-making process. It also can be used to collect individual or collaborative opinions regarding existing situations and help identifying potential conflicts early in the decision-making process and resolving these conflicts through education and compromise later. Some may argue that VGI can be not only efficient but also even better than other data creating methods.

Despite the positive aspects of VGI, there are some challenges or issues faced by VGI. This is caused by the fact that VGI still does not have standards in many aspects despite the recent development. Instead of using scientific standards, VGI tends to be developed towards intuitive, expressive, personal, or artistic geographic techniques. This is not to say that VGI is of no use to the geographic science; it just means that VGI usually does not conform to the protocols of professional practice. For example, VGI does not use the scientific principles of sampling design. It does not use sampling techniques, such as simple random sampling, systematic sampling, and stratified sampling, to generalize useful information from the population. Thus, VGI may not represent the population well. Errors may occur when VGI is used to estimate the entire population information. The estimated figure/information from VGI may not be exactly equal to the true value of the population. For example, VGI has bias towards severe events (Poser and Dransch 2010). The more severe an event and its impacts, the more likely the affected persons report it. Thus, bias is expected for the VGI reported data. The quality of VGI generally increases as the number of contributions increase although a small number of experts may contribute much of the high quality data.

The coverage of VGI varies with different regions in the world—some regions can create or are creating more information or data than others can. This is mainly caused by unevenly accessing to Internet technologies. VGI needs using Internet technology. It needs volunteers to access Internet and GPS-enabled devices. While many regions of the world can access the Internet nowadays, some of regions still have difficulties. Many parts of the world still lack affordable Internet technology and GPS-enabled devices. For example, based on the ICT facts and figures report of the world in 2013 (http://www.itu.int/en/ITU-D/Statistics/Documents/facts/ICTFactsFigures2013-e.pdf), only 16 % of the population in Africa have access to the Internet in 2013, compared to 77 % in the developed world. This may lead to a phenomenon called the digital divide. Another digital divide issue is related to the fact that most web services, where available, can only be used with sufficient knowledge of the English language. This may put those people who don't know English language into the disadvantageous groups for using VGI.

The overall quality of VGI is a much debated issue. The quality of VGI may often be inadequate for practical use or scientific research. Most of the volunteers who provide the spatial information are not geographers or even not scientists, and they lack specialized, formal training in mapping. Comparing with the traditional authoritative geographic information, VGI lacks map specifications, mechanisms or procedures to assure quality. The main mechanism to assure the quality of VGI is called Linus's Law, which states that "given enough eyes, all bugs are shallow" (Raymond 1999). If one volunteer gives wrong information, others will be expected to edit and correct the wrong information. The success of this mechanism depends on others who check the contribution. For example, Wikipedia works fine with this mechanism (Wilkinson and Huberman 2007).

Multiple sources of VGI that ensure vast information availability also make assessing the credibility of information extremely complex. The accuracy of VGI has been compared to reference sources by several studies (e.g. Haklay 2010; Girres and Touya 2010; Cipeluch et al. 2010). For example, Zielstra and Zipf (2010) compared the *OpenStreetMap* to proprietary *TelAtlas MultiNet* data. The results of their analysis show that the VGI of the *OpenStreetMap* project has offered a large amount of data, which is well reflected by the thought of "Citizens as Sensors" (Goodchild 2007). However, they found that the freely available data provided by the *OpenStreetMap* project is not yet a sufficient replacement for the proprietary *TeleAtlas* data for all types of applications, especially a more consistent coverage in rural areas is needed. There is still a very strong heterogeneity in the *OpenStreetMap* data in terms of their completeness. There are significant differences between inner-city and rural areas in terms of the diversity of the freely available data, which can be explained by the presence of more active members on the *OpenStreetMap* project in larger cities. There are also strong differences between large and medium-sized cities in terms of the completeness of the data. For example, the VGI of the *OpenStreetMap* can be a cost-efficient alternative to commercial data sets in the densely populated urban areas of Germany while the coverage of *OpenStreetMap* data in rural areas is too small to be a sophisticated alternative for any application. Different approaches have been proposed to address the quality of VGI. For example, Bishr and Kuhn (2007) proposed trust as a potential proxy measure for the quality of geospatial information.

Given that the creation of its content is complete open, quality may depend on the types of contributors to VGI. A strongly committed expert may contribute high quality VGI while a non-expert such as a passerby contributor may be expected to contribute low quality information. A strongly committed expert may be willing to contribute consistently to make sure the VGI has high quality while a passerby contributor may participate inconsistently thus giving low quality VGI.

Though important, explanations of the motivations of the majority of VGI contributors leave two unanswered questions (Anthony et al. 2005): One is 'what motivates the one-time contributors if not all content comes from committed experts'. The second is 'whether contributor motivations are related to the VGI quality or not'. The experts who contributed much of the VGI content may be motivated by several reasons such as the individual incentives of skill-development, building reputation, and group identity of the community. Some VGI applications have recognized the power of reputation by allowing the interested contributors to become "registered users". It was argued that users may establish a reputation as a registered user instead of an anonymous user. Any contributor who has a strong interest in reputation will register since this is the only way to establish a reputation, while contributors with no interest in reputation may remain anonymous. A different type of incentive for contributors may be the desire to be part of the community. Contributors who identify strongly with the community may participate a lot while contributors who do not identify with the community may likely have low participation. The registered users may contribute the high quality VGI while the anonymous contributors may contribute the low quality VGI. The registered users with high levels of participation may make higher quality contributions through improved skills learned over time (a learning curve) while the anonymous contributors with low levels of participation may make lower quality contributions through their early "career" as contributors. Another possibility may be that users know their contributions are of low quality, thus they do not want to be identified through a registered user name.

A social issue in VGI is the concern that has been raised—how to trust the authenticity and quality of the geospatial information that has been published by individuals? The traditional GIS databases carry implicit quality and authenticity guarantees based on the reputation of the data providers. However, contents from VGI may come from multiple unknown individuals, thus a consumer may want to check the validity of contents before confidently using the data/information. Thus, methods that can help a VGI consumer to check the validity of contents and provide a VGI consumer with a level of trust for contents provided by VGI sites are needed.

Another social issue in VGI is the concerns that VGI may threat to individual privacy. The publishing of contents without control and verification means that users may post data/information that is an invasion of the privacy of others. Personal information posted through VGI may be used for identifying theft, stalking or other harmful privacy invasion intentions. Identifying a VGI publisher may sometimes be difficult when the publisher use anonymous name. However, it is possible to trace an online VGI publisher, thus this kind of anonymity may be only temporary. The identification information from the traced VGI publisher may be used for harmful privacy invasion intentions. In addition, a privacy concern may be arisen when a VGI producer has lots of contents or when the location at which VGI is created is not heavily populated.

VGI has brought many benefits to society, however, it has also brought many challenges and concerns for the copyright. The growth of VGI has caused the controversial issues that relate to intellectual property ownership, especially copyright. While VGI allows users to publish their own data online, some users may publish data belonging to other authors thus violating copyright. Issues like copyright are beginning to be addressed by applying practices from other areas, for example, the open source Creative Commons licensing initiative.

The other issues shared by VGI include data ownership, confidentiality, liability, legal and institutional issues. While VGI has potential to provide spatial data/information for society, it also faces many challenges. For example, the following questions are still waiting to be answered for the future VGI development:

1. How can we engage with the significant volume of spatial information produced by VGI?
2. How can VGI be facilitated and supported with the eventual aim of incorporation into a Spatial Data Infrastructure?
3. How to provide a trusted VGI? Is the traditional identity-based security possible to provide a VGI consumer confidence about the user generated contents without personally knowing who created the contents?
4. How to provide scalable VGI services to a large number of users so that any end-user could be a potential content producer or consumer?
5. How to preserve user privacy/anonymity (including both VGI producers and consumers) if desired by users while VGI becomes ubiquitous and is used by a majority of the population?

Chapter Summary With the development of the Internet, especially Web 2.0, user-generated contents such as "Volunteered Geographic Information (VGI)" have gained a lot of attention and generated a lot of information over the Web. This chapter first introduces the basic information about VGI and presents that VGI (or crowd-sourcing) applications have proven useful for gathering and sharing spatial information. Then it points out that one major problem of VGI is the existence of the heterogeneous semantic problem. A semantic-empowered analysis tool, which is able to automatically integrate and analyze a broad variety of spatial data contents, is needed to provide more utilities for VGI. Later this chapter shows the framework of an interoperable online VGI system based on Geospatial Semantic Web technologies. The left sections in this chapter are focused on introducing (1) information about remotely updating spatial data and (2) algorithms for heterogeneous geospatial ontology integration. Although in IT literatures, algorithms and tools have been proposed to resolve the problem of heterogeneous ontology integration, they are not developed for dealing with spatial data. A Partition-Refinement algorithm for integrating heterogeneous spatial ontology is introduced in this chapter. Compared with the G-Algorithm proposed by Hess et al. (2006), the main advantage of the partition refinement algorithm is that it matches ontology classes and properties based on their structures. After that, a case study of developing an interoperable online VGI system for disaster management in Connecticut is introduced. Finally, this chapter introduces some challenges or issues faced by VGI. The most important issues of VGI are the quality and reliability issues. The other issues include trust, privacy, copyright, data ownership, confidentiality and liability, legal and institutional issues.

References

An Y et al (2005) Inferring complex semantic mappings between relational tables and ontologies from simple correspondences. Lecture notes in computer science for OTM confederated international conferences (Part II), 3761: 1152–1169

Anderson F (2011) How often is Google maps and Google earth updated? http://technicamix.com/2011/10/18/how-often-is-google-maps-and-google-earth-updated/. Accessed 15 May 2015

Anthony D et al (2005) Explaining quality in internet collective goods: zealots and good samaritans in the case of wikipedia. MIT Sloan Technological Innovation, Entrepreneurship and Strategy Seminar website at http://web.mit.edu/iandeseminar/Papers/Fall2005/anthony.pdf. Accessed 25 June 2014

Bahree M (2008) Citizen voices, Forbes Magazine. WWW document. http://www.forbes.com/free_forbes/2008/1208/083.html. Accessed 1 Oct 2013

Bishr M, Kuhn W (2007) Geospatial information bottom-up: a matter of trust and semantics. In: Fabrikant S, Wachowicz M (eds) The European Information Society (2007), pp. 365–387

Castano S et al (2006) Matching ontologies in open networked systems: techniques and applications. In: Spaccapietra S et al (eds) Data semantics V, Lecture notes in computer science. Springer, Berlin, pp 25–63

Cipeluch B et al (2010) Comparison of the accuracy of OpenStreetMap for Ireland with Google maps and bing maps. In: Tate NJ, Fisher PF (eds) Proceedings of 9th international symposium on spatial accuracy: assessment in natural resources and environmental sciences (Accuracy 2010), Leicester, UK, 20-23 July 2010, p 337–340

De Longueville B et al (2009) "OMG, from here, I can see the flames!": a use case of mining location based social networks to acquire spatio-temporal data on forest fires. In: Proceedings of the 2009 international workshop on location based social networks. Seattle, WA, November 2009, 7380

De Longueville B, Luraschi G, Smits P et al (2010) Citizens as sensors for natural hazards: a VGI integration workflow. Geomatica 64:41–59

De Longueville B, Annoni A, Schade S et al (2011) Digital earth's nervous system for crisis events: real-time sensor web enablement of volunteered geographic information. Inter J Digital Earth 3:242–259

De Rubeis V et al (2009) Web based macroseismic survey: fast information exchange and elaboration of seismic intensity effects in Italy. In: Proceeding of the 6th International ISCRAM conference, Gothenburg, Sweden, May 2009. WWW document. http://www.iscram.org/ISCRAM2009/papers/. Accessed 1 Oct 2013

Gao H et al (2011) Promoting coordination for disaster relief—from crowdsourcing to coordination. Lect Notes Comput Sci 6589:197–204

Girres JF, Touya G (2010) Quality assessment of the French OpenStreetMap dataset. Trans GIS 14:435–459

Goodchild MF (2007) Citizens as sensors: the world of volunteered geography. GeoJournal 69:211–221

Goodchild MF, Glennon JA (2010) Crowdsourcing geographic information for disaster response: a research frontier. Inter J Digital Earth 3:231–241

Haklay M (2010) How good is volunteered geographical information? a comparative study of OpenStreetMap and ordnance survey datasets. Environ Plan B: Plan Des 37:682–703

Hess GN et al (2006) An algorithm and implementation for GeoOntologies integration [online]. In: GEOINFO 2006, Campos do Jordão, Brazil. http://www.geoinfo.info/geoinfo2006/papers/p46.pdf. Accessed 6 May 2009

Hughes AL, Palen L (2009) Twitter adoption and use in mass convergence and emergency events. In: Proceedings of the 6th international ISCRAM conference, Gothenburg, Sweden, May 2009. WWW document. http://www.iscram.org/ISCRAM2009/papers/. Accessed 1 Oct 2013

Jha S et al (2002) Efficient type matching. In: Proceedings of FOSSACS'02, foundations of software science and computation structures. Lect Notes Comput Sci 2303:187–204

Laituri M, Kodrich K (2008) On line disaster response community: people as sensors of high magnitude disasters using internet GIS. Sensors 8:3037–3055

Morrow N et al (2011) Independent evaluation of the ushahidi haiti project. Tech. rep. The UHP Independent Evaluation Team. WWW document. http://ggs684.pbworks.com/w/file/fetch/60819963/1282.pdf. Accessed 1 Oct 2013

Palen L, Vieweg S, Liu SB et al (2009) Crisis in a networked world: features of computer-mediated communication in the April 16, 2007, Virginia Tech event. Soc Sci Comput Rev 27:467–480

Peng ZR, Zhang C (2004) The roles of geography markup language, scalable vector graphics, and web feature service specifications in the development of internet geographic information systems. J Geogr Syst 6:95–116

Poser K, Dransch D (2010) Volunteered geographic information for disaster management with application to rapid flood damage estimation. Geomatica 64:89–98

Raymond ES (1999) The cathedral and the bazaar: Musings on Linux and Open Source by an accidental revolutionary. O'Reilly, Beijing

Shvaiko P, Euzenat J (2005) A survey of schema-based matching approaches. Lect Notes Comput Sci: J Data Semant IV(3730):146–171

Stoilos G et al (2005) A string metric for ontology alignment. Lect Notes Comput Sci Inter Semant Web Conf 3729:624–637

Wikimapia.org (2014) "Places total" statistics page. http://wikimapia.org/#lang=en&lat=41.808000&lon=-72.251000&z=12&m=b&show=/stats/action_stats/?fstat=101&period=1&year=2014&month=6. http://en.wikipedia.org/wiki/WikiMapia. Accessed on 23 Jan 2014

Wilkinson DM, Huberman BA (2007) Assessing the value of cooperation in Wikipedia. http://arxiv.org/abs/cs/0702140. Accessed 15 May 2015

Zielstra D, Zipf A (2010) A Comparative Study of Proprietary Geodata and Volunteered Geographic Information for Germany. 13th AGILE International Conference on Geographic Information Science 2010, Guimarães, Portugal

Zook M, Graham M, Shelton T et al (2010) Volunteered geographic information and crowdsourcing disaster relief: a case study of the Haitian earthquake. World Med Health Policy 2:7–33

Zhang C, Li W (2005) The roles of Web Feature and Web Map services in real time geospatial data sharing for time-critical applications. Carto Geogr Info Sci 32:269–283

Zhang C, Li W, Peng ZR et al (2003) GML-based Interoperable geographical database. Cartography 32:1–16

Zhang C, Li W, Zhao T (2007) Geospatial data sharing based on geospatial semantic web technologies. J Spat Sci 52:11–25

Zhang C, Zhao T, Li W et al (2010a) Towards logic-based geospatial feature discovery and integration using Web Feature Service and Geospatial Semantic Web. Inter J Geogr Info Sci 24:903–923

Zhang C, Zhao T, Li W (2010b) A framework for geospatial semantic web based spatial decision support system. Inter J Digital Earth 3:111–134

Zhang C, Zhao T, Li W (2010c) Automatic search of geospatial features for disaster and emergency management. Inter J Appl Earth Obs Geoinfo 12:409–418

Zhao T, Zhang C, Wei M et al (2008) Ontology-based geospatial data query and integration. In: Geographic Information Science. Lecture Notes Computer Science 5266:370–392

Chapter 6
Applications of Geospatial Semantic Web

6.1 Background

Geospatial semantic web has the potential to improve the intelligence, flexibility, and usability of geospatial applications. In this chapter, we describe several applicable areas of geospatial semantic web and their prototype implementations. These areas include disaster and emergency response systems (Zhang et al. 2010a), volunteered geospatial information systems (Zhang et al. 2014, Zhang et al. 2013), transit network information systems (Zhang et al. 2010b), and spatial decision support systems (Zhang et al. 2010c). The protocols, workflows, and algorithms used in the prototypes are covered in previous chapters so that this chapter focuses on the software libraries and the programming aspects of the implementations.

- The implementations of these geospatial applications include the following components:

  1. geospatial data sources for map layers and geometries of spatial objects,
  2. application logic for query processing, data transformation, and integration,
  3. user interfaces for rendering map layers, geometries, and attributes of spatial objects, and for accepting user queries and manipulations.

With geospatial semantic web, these components can be weaved together to produce flexible query interfaces, fine-grained data access, and short response time. To achieve these goals, we need to overcome several obstacles. For example, data distributed in many sources and applications need to provide a uniform access to these sources in order to retrieve and integrate data from the multiple sources in a transparent way. An emergency response system may require data from USGS map servers, local GIS centers, and weather services. Using ontology definitions, we can define data from these sources in a single representation. We can use ontology namespaces to provide abstractions for data from each source. Ontology query system can import these namespaces into a single environment to enable uniform data access. We also can use ontology classes and properties to define mappings from the original data definitions to the application data definitions in order to bridge

© Springer International Publishing Switzerland 2015
C. Zhang et al., *Geospatial Semantic Web*, DOI 10.1007/978-3-319-17801-1_6

the semantic gap between source data sets. The application logic can interact with ontology data instead of the data of the original sources. User interfaces can utilize the standard query protocol for access of geospatial ontology—GeoSPARQL can be used to access the original data sources as ontology data. Ontology representation brings the benefit of ontology reasoning so that indirect information can be extracted from the direct data knowledge. Spatial objects can be retrieved based on query predicates that correspond to high-level semantics of a user domain. In general, the geospatial semantic web serves as the glue to bridge the semantic gap between data sources and can provide fine-grained access to heterogeneous spatial data. Flexibility of geospatial semantic web does bring significant runtime overhead, which impacts the responding time of geospatial applications. We show how this overhead can be overcome in application implementation in this chapter.

6.1.1 Relevant Software Tools and Libraries

Apache Tomcat (tomcat.apache.org) is a Java-based Servlet container for creating web services. It provides an execution environment for deploying web services that are essential to the prototypes described in this chapter. In the prototypes, both the data sources and the application logic components are implemented as web services. These components communicate remotely over the HTTP protocol so that they can be hosted in multiple servlet containers located in the same or different physical servers. Apache Tomcat also maintains the life cycle and security of web services.

GeoServer (geoserver.sourceforge.net) is an implementation of several OGC web service protocols: WFS, WMS, and WCS. As a web service application, GeoServer can be deployed in an Apache Tomcat servlet container. Once deployed, the OGC web services of a GeoServer are available as HTTP calls. GeoServer accepts multiple forms of data stores such as Shapefiles and geospatial databases. By providing geospatial data sources as web services, GeoServer effectively acts as a standard data source that provides more fine-grained and convenient access to spatial objects than Shapefiles or geospatial databases do.

Jena (jena.sourceforge.net) is a semantic web framework library, which provides a large number of functionalities for ontology data manipulation. It is used in the prototypes for generating ontology model from other data sources, for outputting ontology classes, properties, and instances to data stores, and for providing ontology reasoning capability.

Joseki (www.joseki.org) is an ontology web service, which provides SPARQL query endpoints. The prototypes in this chapter utilize Joseki endpoints to conduct spatial and non-spatial data queries that are represented as ontology instances through SPARQL queries. The ontology data stores for Joseki endpoints are generated through the Jena library. The internal implementation of Joseki is also based on Jena.

OpenLayers (openlayers.org) is a JavaScript-based geospatial library for rendering map layers and spatial objects on a browser interface and for accessing geospatial data in web services, which include OGC web services provided by GeoServers and map services, such as those provided by Google Map, Bing Map, and Open Street Map. OpenLayers supports web service calls to OGC web services through its AJAX implementation to retrieve maps and spatial features. Openlayers includes two types of layers: map layers for raster map images and vector layers for vector features. Openlayers provides common control widgets for layer manipulation such as pan, zoom, and switching of layers. More fine-grained controls in Openlayers are implemented by event handlers that respond to user events from mouses and keyboards. Using event handlers, applications can select, drag, and edit features in vector layers. Additional controllers can be implemented to dynamically add features to or remove features from vector layers. Finally, the features in vector layers can be made persistent in their originating WFS services.

In the following sections, we introduce several applications of Geospatial Semantic Web, including a NLP-based geospatial query interface, a volunteered geospatial information system, and a geospatial interface for transit network.

6.2   NLP (Natural Language Processing) Based Geospatial Query Interface

6.2.1   Motivation

Emergency and disaster responses require timely access to geospatial data such as airports, road networks, and town boundaries. However, the first responders to emergencies and disasters often lack proper training to use geospatial applications. Ontology-based interface presents challenges to users who are not familiar with semantic web. It may be daunting for them to formulate SPARQL queries to extract geospatial features from the ontology-based knowledge base and web services. A more intuitive interface can greatly enhance the usability of geospatial applications in emergency situations. Such an interface can be based on processing natural language queries that do not require prior knowledge in ontology or SPARQL. The NLP (Natural Language Processing) based interface takes queries of short text sentences as input and translates them to ontology query requests sent to multiple sources with the help of the ontology knowledge base. The interface takes advantage of semantic interpretation of the natural language queries and exposes WFS services directly to non-expert users.

Through OWL ontologies in the ontology server, the NLP interface provides richer semantic specifications for WFS services; users can implicitly identify their desired *feature types* and *feature properties* and properly interpret the query results. Thus, it is possible to perform an intelligent content-based search of WFS services at semantic level and to query geospatial features from semantically heterogeneous sources.

6.2.2   The Prototype

The NLP interface prototype provides an intuitive and easy-to-use interface to allow users to query the ontology-based knowledge base using a natural language.

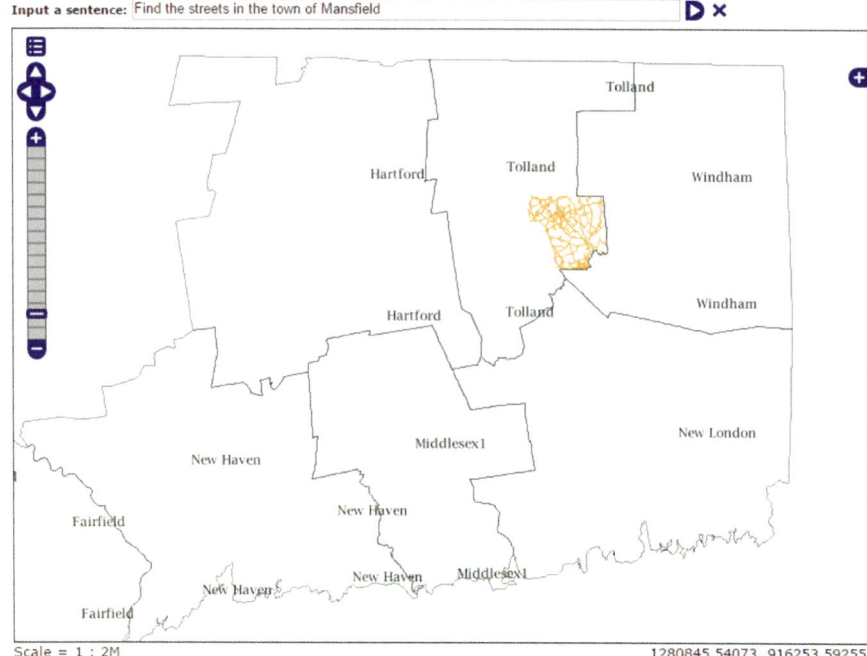

Input a sentence: Find the streets in the town of Mansfield

Fig. 6.1 The natural language interface of the implemented prototype

For the interface prototype, we use several types of spatial features in the state of Connecticut, which include maps of town, county, and state boundaries, road and street networks, and airports. Figure 6.1 shows the natural language interface of the implemented prototype to query street features in Mansfield by using natural language "Find the streets in the town of Mansfield". The prototype parses the natural language queries and translates them to formal SPARQL queries sent to an ontology data server. The returned streets are those within the town of Mansfield. In fact, the results come from analysis of the two datasets in WFS services: the Connecticut street dataset and the town boundary dataset. Retrieving street features is slower because of the large number of line segments in streets.

The prototype accepts short sentence queries that specify the subject such as:

1. Find all the airports.
2. Find the major roads.
3. Find all the streets.
4. Find all the towns

It also accepts queries with qualifiers to restrict the location of the queried subject such as:

5. Find the major roads across the town of Mansfield.
6. Find the major roads within the Hartford County.
7. Find the streets in the town of Mansfield,

   8. Find the airports in the town of Groton,
   9. Find the airports in the Hartford County,
  10. Find the towns of the Tolland County.

The former type of queries selects all the objects of a feature type while the latter type of queries only selects objects of a feature type subject to a spatial join filter.

6.2.3   *Implementation*

The prototype allows emergency responders to automatically search geospatial features, such as roads, town boundary, and airports in Connecticut served by the different WFS services, from multiple semantically heterogeneous data sources.

   There are many challenges in the development of the natural language interface. The ambiguity and complexity of a natural language make it difficult to develop an algorithm to parse and capture the correct semantics in the natural language queries. We need to translate the natural language queries to formal SPARQL queries and correctly map the terms in the queries to terms in the ontology knowledge base. A choice is to use the application and domain specific ontologies to produce the lexicon and grammar rules for parsing a user's input. The grammar is constructed programmatically from the application and domain ontologies to create the verbs, nouns, and prepositional phrases corresponding to the classes and the properties in the ontologies.

   A NLP parser is used to extract a syntax tree from each text query. The synonyms for all vocabularies are provided by *WordNet*. From the syntax tree, the prototype extracts a sequence of the main word categories *Noun (N), Verb (V), Preposition (P), Wh-Word (Q), and Conjunction (C)*. The prototype generates a query skeleton from the extracted word categories and then matches the query skeleton with the synonym-enhanced triples in the ontologies. The matching is controlled by domain and range definitions of the ontology properties. After identifying all possible triples in the query skeleton, the prototype combines those triples that share ontology resources to generate the *query triples* for the SPARQL queries sent to ontology servers.

   The *lexicon* used in the framework is composed of three sources: (1) ontology entities in an ontology server, including ontology classes (concepts), ontology properties (relations), and ontology instances (individuals), are used to limit the ambiguities and errors in the natural language interactions; (2) general dictionaries, such as *WordNet*, are used to enlarge the vocabulary of the ontology and help mapping user vocabulary to ontology vocabulary; (3) application specific synonyms, such as user-defined synonymy words, are used to define application jargons and abbreviations. The *lexicon* can be updated in an automatic way from the ontology server for the specific application and domain of use. The semantic values (classes) that inherit from the ontologies used by the applications can be inserted to the words in the *lexicon*.

   The implementation of the NLP interface depends on an underlying ontology-based query interface to connect to the data sources.

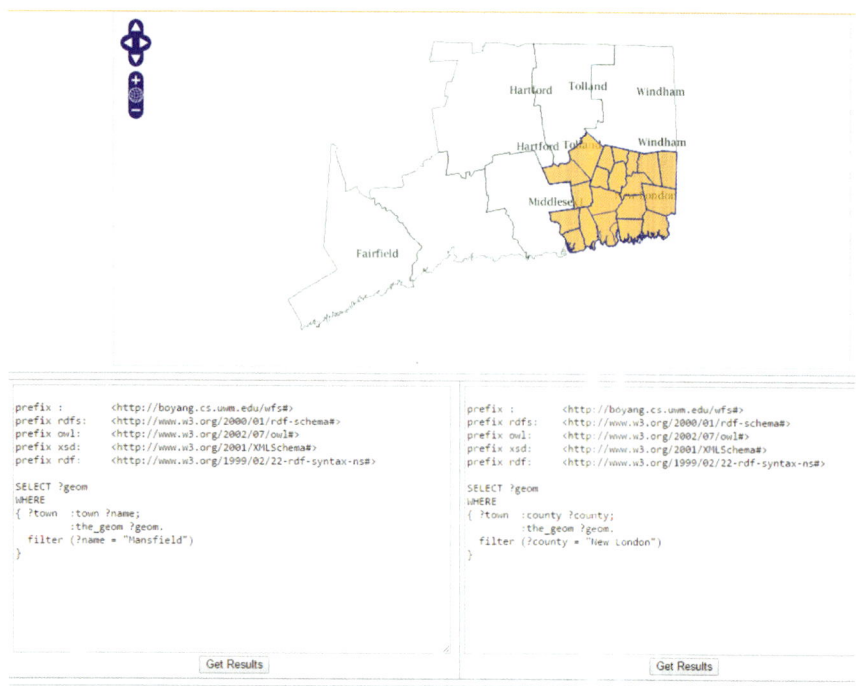

Fig. 6.2 Underlying SPARQL query interface

Figure 6.2 illustrates an example of the ontology query interface, where the lower left box is a query for the town of Mansfield and the lower right box is a query for the towns in New London county. The map in Fig. 6.2 shows the results of the right query, which are the geometries of the towns in "New London" county.

To store geospatial data and answer geospatial queries, we utilize GeoServer's WFS services. A spatial ontology server based on Joseki is created to answer ontology queries in SPARQL forms. Jena library is used to store ontology definitions, to analyze their consistency, and to infer knowledge needed for sending WFS requests. The clients of the ontology-based and natural language-based user interface are developed using OpenLayers, which renders the query results as map layers. Stanford Parser is used as the natural language parser. Apache Tomcat is used to host our server instances, which include GeoServer, Joseki ontology server, and NLP server.

6.2.4   Web Service Preprocessing

When users send natural language queries, the NLP interface prototype converts the queries to ontology queries using SPARQL. The SPARQL queries are then rewritten to WFS *getFeature* requests via ontology reasoning or inference rules if applicable. To discover which web services contain the geospatial data to answer user queries, we need to check a list of known WFS services. Instead of performing this check for each user query, we preprocess the WFS services to match the feature types in the services with local ontology so that user queries do not need to require repeated lookup of the meta data of WFS services.

The prototype preprocesses the web service metadata and creates connections between the web services and the ontology definitions. The main procedures for creating the connections are the following: The prototype takes a list of known WFS services and sends a *getCapabilities* request to each of the services. The returned feature data is parsed to obtain the list of feature types in each web service, plus the name, the bounding box, and the description of each service. For each feature type, a *describeFeatureType* request is sent to obtain their properties and types including the geometry types. Next, the prototype compares the feature names and their descriptions with the local ontology to map the features to the ontology classes. The process is repeated for the local ontology properties. One ontology class is matched to one WFS feature type. Each ontology class simulates a WFS feature type so that the class includes all the feature properties as ontology properties. A companion ontology instance is generated for each WFS feature type to include other feature information such as server URL, bounding box, description, geometry type, and a list of property names and their types. Based on the mapping between WFS features and domain/application ontologies, subclass relations to the generated ontology classes are added and the equivalent ontology properties are merged. The built-in reasoning algorithm of Jena is used to infer feature instances (i.e. route segments and stops) that users would like to retrieve. Jena is also used indirectly in the Joseki server for ontology inference.

6.3 Volunteered Geospatial Information System

6.3.1 Motivation

New Haven, a coastal city in Connecticut, USA, has endured many storm disasters in the past. The tropical storms in 2011 exposed the lack of an efficient information system for responding to and managing the storm disasters. To provide timely, accurate, and up-to-date information for better disaster response and management, a prototype is implemented to allow disaster responders and volunteers to share and update geospatial data over the network. The prototype also addresses the semantic heterogeneity problems of geospatial data.

The main goal of the system is to facilitate real-time sharing of the existing legacy geospatial data and the new updated information among different organizations, such as town responders, utility companies, medical departments, and police departments. The disaster responders and managers, therefore, can work together and coordinate seamlessly in the event of a storm disaster. The prototype should also be machine-readable so that computers can understand the content of the geospatial databases, and thus the required information for disaster response can be readily retrieved from multiple data sources.

6.3.2   The Prototype

The prototype interface is a dynamic HTML page embedded with about a thousand lines of JavaScript code that calls *OpenLayers* library to query/update/render geospatial features through WFS/WMS services. The web service backend is an instance of *GeoServer* that provides WFS/WMS services for several types of geospatial features. As shown in Fig. 6.3, the interface includes an interactive map with several control buttons and a form for ontology-based query. The buttons "*Sel streets*" and "*Sel roads*" allow users to select street or road features on the map using mouse. After selecting some features, the users can click on the buttons "*Block streets*" or "*Block roads*" to save the selected streets or roads as being blocked by the hurricane storm event. If the users find out that they have made a mistake, they can change the "*blocked*" attributes to "*unblocked*" status by selecting the "*blocked*" streets or roads and clicking on the "*Unblock streets*" or "*Unblock roads*" buttons.

Other than directly selecting streets or roads, users can also click on the "*Sel places*" button to enable a control that allows users to select point features such as schools and bridges from the map using mouse. After some places are selected, the interface client can display the attribute values of the selected places below the map along with two hyperlinks that allow users to select the streets and roads nearby the selected places. Figure 6.4 shows two selected places in blue circles and their attributes are shown below the map, which indicate that one of the places is a railroad station and the other place is a high school.

The displayed feature attributes are transformed through ontology classes from the original attributes stored in the data source. For example, the attribute for rep-

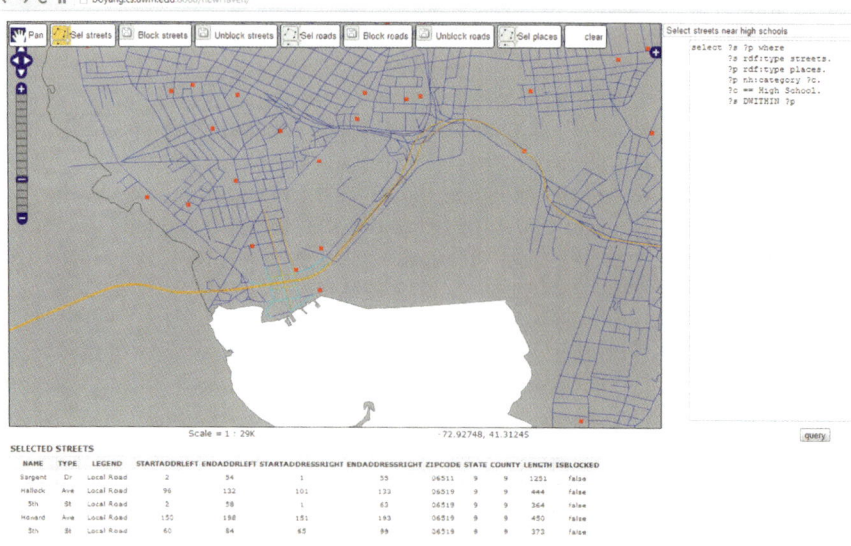

Fig. 6.3 The client interface of the implemented prototype

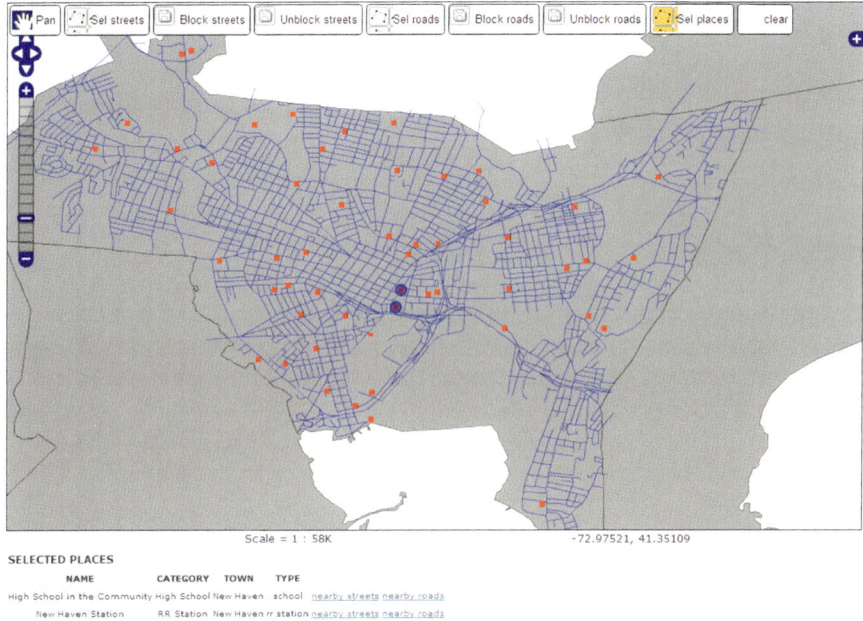

Fig. 6.4 Attributes and hyperlinks of the two selected places

resenting whether a street is blocked is called "*Blocked*" while a similar attribute for road feature is called "*Obstructed*". We create a mapping from these attribute names to an ontology property so that they are all accessed as "*isBlocked*" property.

The interface also includes a query form to allow users to select features using ontology-based query statements. The ontology query can be used to select features using more fine-grained criteria that are difficult to express using the graphic interface. For example, we may want to find out the elementary schools and streets near the road "*Connecticut Tpke*". We can write this query using a SPARQL-like query statement to select features of the "*streets*", "*roads*", and "*places*" types so that the "*category*" of the places is "*Elementary school*", the road name is "*Connecticut Tpke*" and the streets, the roads, and the places are near each other (i.e. their distances are within a certain threshold).

The prototype can not only dynamically update new sources of information but also integrate heterogeneous information at the semantic level. It, therefore, can be used by users including disaster responders and the local residents to report information about events and damages in their neighborhood and update the related geospatial information over the web. Users of the prototype need only to understand the definitions for general concepts of streets, roads, and places to be able to query and update geospatial information; they don't need to understand the underlying definitions of the geospatial data, which may be different across regions or data sources. As long as the definitions of the original geospatial data are properly mapped to the ontology-based definitions, the ontology-based queries can be understood by the prototype. For example, to query the highways with high schools nearby, users only need to remember to use ontology property *nh:category* to specify the type of the place as high school, use the property *nh:legend* to specify the type of the streets as highway, and use the property *DWITHIN* to specify geospatial relationships between the queried objects.

6.3.3 Implementation

The prototype uses a Geoserver instance running in a Tomcat servlet container to provide the backend WFS and WMS services, and a Web frontend using Open-Layers library to provide map rendering. The core logic of the prototype is implemented in JavaScript that converts SPARQL queries to WFS requests, which are sent as AJAX calls to the backend geospatial web services. The results of the AJAX calls are joined if necessary to synthesize the map layers to be displayed in a Web browser.

The prototype uses several layers of map data for New Haven, Connecticut and its surrounding areas. The map layers include *streets*, *roads*, *places*, and *city boundaries*. Both *street* and *road* features are line geometries, though *street* geometries are mostly shorter than the *road* geometries and *street* features have more detailed attributes than the *road* features. The *places* features are point geometries for locations such as schools, bridges, and rail road stations.

For our experiments, we create only a few of ontology classes to demonstrate the usability of the prototype. Figure 6.5 illustrates the ontology classes created for the prototype. The ontology class *Line* is the parent ontology class of *Streets* and *Roads* while *Point* is the parent ontology class of *Places*. In the future, additional classes can be introduced to the hierarchy without any change to the implementation.

The *Streets* ontology class is mapped to the *streets* feature type in the backend web feature services, while *Roads* and *Places* ontology classes are mapped to the *roads* and *places* feature types, respectively. The properties of the ontology classes are mapped to the attributes of the feature types. For example, the following is the mapping between the properties of the *Roads* ontology class and the *nh:roads* feature type in OGC web feature services:

Fig. 6.5 Ontology classes created for the prototype

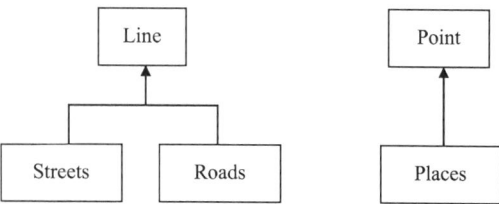

Attributes of *roads* feature type	Properties of *Roads* ontology class
'OBJECTID'	'id'
'FULLNAME'	'name'
'RTTYP'	'type'
'Shape_Leng'	'length'
'Blocked'	'isBlocked'

Similar mappings are defined for values of some attributes to provide more readable query results. For example, below is the mapping between the values of the attribute '*RTTYP*' in the original data source and the values of the property '*type*' in the *Roads* ontology class:

Values of the attribute '*RTTYP*'	Values of the property '*type*'
'C'	'County'
'I'	'Interstate'
'M'	'Common name'
'O'	'Other'
'S'	'State recognized'
'U'	'US'

By using ontology classes, the implemented prototype can allow different stakeholder groups, such as local authorities, relief specialists, nongovernmental agencies, disaster responders, and the local residents, to understand the vocabulary of the original data. Thus, it can deliver more meaningful geospatial information, which could be more useful in the disaster response and management.

In addition to the *datatype* properties mapped from Web feature attributes, each ontology class includes object properties, such as *DWITHIN*, to allow users to specify spatial queries in the form of ontology constraints. For example, the following ontology query retrieves streets near any high school:

select *?s ?p* **where**

  ?s *rdf:type* *streets.*

  ?p *nh:category* *?c.*

  ?c *=* *High School.*

  ?s *DWITHIN* *?p*

where the variable?s and?p are geospatial features that are displayed in the map in Fig. 6.6.

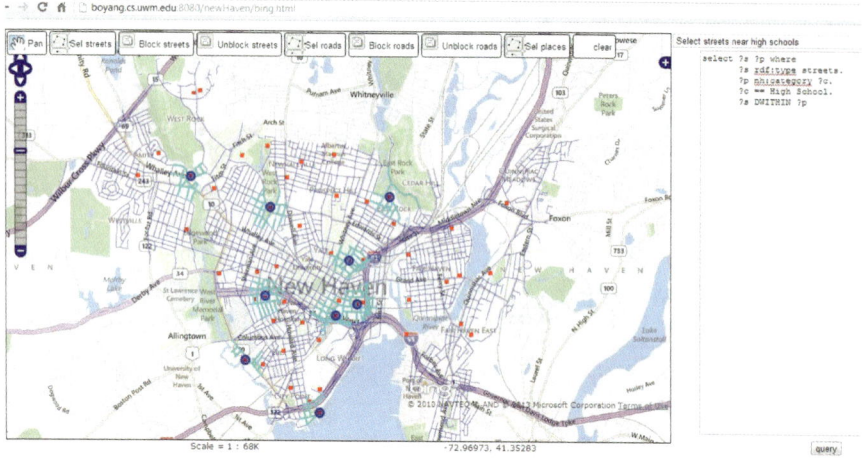

Fig. 6.6 Ontology query results of streets near high schools in New Haven, Connecticut

The constraints *?s rdf:type streets* and *?p rdf:type places* are triples of the form **subject predicate object,** which specify that the variable *?s* must be assigned instances of the *streets* type and *?p* must be assigned *places* type. The constraint *?p nh:category?c* creates a new variable *?c*, which is set to equal to *High School* in the constraint *?c = High School*. The predicate *nh:category* is based on the ontology property category of the *Streets* class and *nh:* is a namespace prefix to distinguish the property from other types of predicates. Finally, the constraint *?s DWITHIN?p* says that the retrieved streets *?s* must be near the retrieved places *?p*. The predicate *DWITHIN* corresponds to an ontology property that is not mapped to attributes since it relates two geospatial objects by the distance between their geometries. In fact, there is no explicit representation of the relationship because it would be too inefficient to pre-compute and store the relationships while only a small fraction of them will be needed.

An additional benefit of ontology based query is that users do not need to always specify the types of the queried objects. For example, the following query returns any geospatial objects by the name of *Lawrence*:

$$\textbf{\textit{select}} \quad \textit{?s} \quad \textbf{\textit{where}}$$

$$\textit{?s} \quad \textit{nh:name} \quad \textit{?name.}$$

$$\textit{?name} \quad \sim \quad \textit{Lawrence*}$$

where ~ is a predicate for "*LIKE*" relationship to match names such as *Lawrence St*. The three ontology classes we defined all have the property "*name*". The above query is in fact translated into three separate WFS requests to retrieve features of the Web feature types: *street*, *road*, and *places*. Each of the WFS requests includes a filter with a different name attribute: FENAME for street, FULLNAME for road, and NAME for places.

6.3.4   Experience

Although the results of the prototype show some promises towards semantic interoperability for geospatial web services, many issues, such as fault/error handling and security issues, still need further research to implement a fully workable system for the real world applications. Here we address the performance issue in the following paragraphs.

Performance is an important issue that needs to be solved in the future. Although the prototype only involves data from one coastal city in Connecticut—New Heaven, some layers contain several thousands of geospatial features, for example, the street layer including almost four thousands of geospatial features. Many more geospatial features will be involved if we implement the prototype for the whole state Connecticut. Even for the New Heaven data sets, the system becomes slow in updating the street layer data. This is because the WFS needs to transport text-based GML data over the network. As we mentioned in the previous book chapter, when GML-coded geospatial data are transported, all the markup elements that describe geospatial and non-spatial features, geometry, and geospatial reference systems of the data are also transported to the recipient. This is important for data interoperability. However, this greatly slows down the performance of the system. Compared with some binary GIS data formats, the size of GML data files is large. Large file sizes may hinder the use of GML files as a means of data transport over the Internet. Although solutions, such as using compression or sending the GML data to the client in stages or progressively, have been proposed in literature (Peng and Zhang 2004), issues still exist for igniting the performance.

To evaluate the performance of this approach, we compared it with an optimized implementation of the ontology query engine described in Zhao et al. (2014). The optimized implementation extends the Jena ontology framework with some optimization methods, which include parallelization, runtime indexing of spatial objects, and caching of geometries parsed from ontology literals, in order to improve the performance of answering GeoSPARQL queries. Here we use an example query of *"selecting the streets nearby each school in New Haven, CT."* for the comparison purpose. Using the same New Haven data set with 54 schools and 3449 streets, Zhao et al. (2014) reduced the runtime of answering the example query from 4086 ms (or 4 s) with the original Jena query engine (*cf.* jena.apache.org) to 703 ms with one thread to 432 ms with four threads. In this comparison study, while the runtime of the current approach took 5700 ms when the query was answered for the first time, the processing time dropped to 330 ms on average when we repeated the query. Note that the runtime we measured for the current approach also included the time spent on rendering and displaying the spatial objects on map layers. We achieved a performance gain for the repeated queries by taking advantages of caching the features retrieved from WFS servers. We cached features based on the WFS requests so that an ontology query could help to improve the performance of a subsequent query if their translations include the same WFS requests. For example, if we have two queries— (1) *"select streets nearby each **high** school"* and (2) *"select streets nearby each **middle** school"*, we need to fetch the street features from WFS servers and cache them locally to answer the first query. To answer the second query, we also need the street features, but we can read them from the local cache because the WFS requests translated from both queries contain the same request for the street features. Thus, through caching retrieved features, the processing runtime can be greatly reduced.

The current approach is faster than the optimized ontology query engine, despite the fact that the runtime of the previous method is measured on a Java implementation with direct access to the ontology data while the current approach is based on remote network access to WFS servers using a JavaScript implementation running as a single thread in a browser. We overcame the overhead of encoding, decoding, and transporting a large number of spatial objects to answer the query through caching. To answer the example query, all 3449 street objects are fetched from the WFS server. This step accounts for the majority of the runtime when the query is processed the first time. The retrieval of school objects and the spatial join process consume less than 200 ms. Since we have cached the features fetched from servers, there is no need to send requests to WFS servers again if the same features are needed in subsequent queries.

In addition to caching, we can reduce the number of spatial objects retrieved from WFS servers through several other alternatives. These other alternatives include:

1. Sending the spatial queries to servers. For the query to select streets nearby each school, we can use the retrieved point geometries of the schools to formulate spatial queries and send them to WFS servers to retrieve the street objects. This should greatly reduce the number of street objects as well. The drawback, however, is that if the number of schools is large, then the spatial query sent to WFS servers may exceed the maximum size of HTTP calls. Under this situation, we can break down the call into several calls to keep the size below the limit. However, multiple calls to WFS servers may cause additional delay. In addition, spatial queries like this may result in high runtime costs to WFS servers, thus reducing the overall performance.

2. Applying spatial joins of the query to WFS servers. For the query that selects nearby streets of each school, we can add a cross-layer extension to the GeoServer implementation of WFS servers to answer a query so as to filter street objects using the distance to each school:

```
<wfs:Query typeName="nh:streets">

  <ogc:Filter>

    <ogc:DWithin>

      <ogc:PropertyName>the_geom</ogc:PropertyName>
```

```
<ogc:Function name="collectGeometries">

    <ogc:Function name="queryCollection">

        <ogc:Literal>nh:places</ogc:Literal>

        <ogc:Literal>the_geom</ogc:Literal>

        <ogc:Literal>INCLUDE</ogc:Literal>

    </ogc:Function>

</ogc:Function>

<ogc:Distance units="meter">100</ogc:Distance>

</ogc:DWithin>

</ogc:Filter>

</wfs:Query>
```

The drawback of this approach is that we have to modify the query rewriting algorithm to extract the spatial joins that are sent to WFS servers. However, there is no guarantee that the spatial joins will result in performance improvement for all types of queries. In addition, we can only apply spatial joints to the WFS layers of the same server. However, we may often use multiple WSF servers for a real-world application.

Performance improvement for geospatial ontology query was also investigated by Dellis and Paliouras (2005), who used R-tree to index spatial data separated from the non-spatial ontology data. Their experiments showed that the runtime for range queries and K-Nearest Neighborhood queries that only involve spatial attributes was dramatically reduced and the approach was very scalable. The performance boost is not surprising since they essentially performed queries on an indexed spatial database. For queries that involve both spatial and non-spatial attributes, their experiments also showed performance gain. For our case, indexing the spatial objects using R-tree is not necessary at the server side, because we rely on the WFS server implementation—GeoServer, which already supports indexing. Thus, we already took advantage of spatial indexing to answer range queries. However, we could apply spatial indexing at the client side to the cached spatial objects. This should be a research direction left for future work.

6.4   Geospatial Interface for Transit Network

6.4.1   Motivation

Real-time traffic applications need real-time access and exchange of the most up-to-date feature-level data. OGC WFS services facilitate feature-level spatial data sharing over the web in real time. However, WFS services only emphasize technical data interoperability via standard interfaces and cannot resolve semantic heterogeneity problems in spatial data sharing. With currently implemented WFS services, it is only possible to search and access geospatial data by keywords in metadata. In addition, the lack of explicit semantics in the WFS description proves to be a major limitation for automatic discovery of geospatial features and composition of WFSs. To overcome these limitations, we developed a solution for searching, discovering, and composing semantically heterogeneous transportation spatial data at the feature level from different sources over the web through providing semantic specifications of WFS services. Geospatial semantic web technologies, such as Description Logic-based reasoning, inference rules, and OWL (Web Ontology Language) ontologies, were used to support geospatial feature data interoperability at the semantic level. Algorithms for discovering and composing geospatial features and services were developed for this goal.

6.4.2   The Prototype

A prototype was implemented to search and access semantically heterogeneous geospatial features in transit data. With semantic descriptions of WFS services, the prototype allows users to discover, compose, and synthesize semantically heterogeneous geospatial features. Figure 6.7 shows the results of the automatically dis-

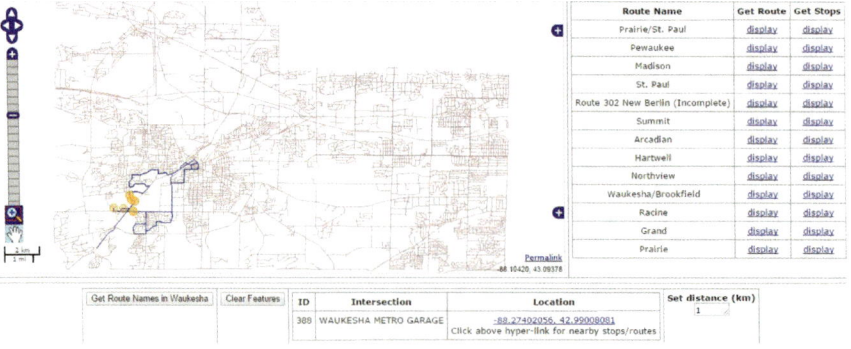

Fig. 6.7 Bus roads and bus stops located within 1 km of the bus stop in Fox Run Shopping Center of downtown Waukesha

covered and composed bus roads and bus stops located within 1 km of the bus stop in the Fox Run Shopping Center of the downtown Waukesha. Using the prototype, users can get all the routes in the downtown Waukesha by clicking the button "Get Route Names in Waukesha". Users can then query an individual route by clicking the corresponding "display" link in the column of "Get Route", and also can query all bus stops related to a route by clicking the corresponding "display" link in the column of "Get Stops". Particularly, users can query and compose bus roads and bus stops within a certain distance (default is 1 km) of the selected bus stop by clicking the hyper-link of the location (user should first click to select a bus stop geometry).

While semantically enhanced WFSs can help to automate the discovery and integration of geospatial features, they can introduce significant runtime overhead with ontology reasoning when there are a large number of ontology instances. This may impede the real-time access and exchange of the geospatial feature data. The performance of TBox reasoning (reasoning about concepts) is generally acceptable. However, ABox reasoning (reasoning about individuals) can be quite slow for some realistic sets of spatial data. Because the implemented prototype manipulates thousands of feature instances, we can afford converting all these feature instances into ontology instances and then perform geospatial reasoning. The performance of the prototype for user queries and composition of bus routes and stops is comparable to that of direct access to WFS servers. However, the transportation network in a larger city may contain tens of thousands to millions of feature instances, which could make it impractical to convert WFS features into ontology instances and then conduct reasoning. In fact, we decided not to convert the street network of Waukesha to ontology instances in our implementation because it would incur noticeable delay in querying. We chose to store the street network in a WMS server and relay WMS queries from our ontology server to retrieve street maps instead. Thus, the challenge is how to improve the efficiency of geospatial reasoning so that it is still practical to allow real-time combination of geospatial features from various sources. In this case, we need to avoid converting all geospatial features into ontology instances. Instead, we need a query pre-processor to translate the A-Box queries into some sub-queries, which will be sent to spatial or non-spatial data sources such as WFS servers or databases to get the needed data. This method can greatly enhance query efficiency, though we may lose some reasoning power considering that we cannot apply reasoners to geospatial features directly.

6.4.3   Implementation

Figure 6.8 illustrates the architecture of the implemented prototype, which consists of an instance of Geoserver to provide WFS and WMS services backed by a ShapeFile data store, an Apache Tomcat Servlet container to host the GeoServer and spatial ontology server, and a spatial ontology server based on Joseki to answer SPARQL ontology queries. The ontology server uses a domain ontology for spatial

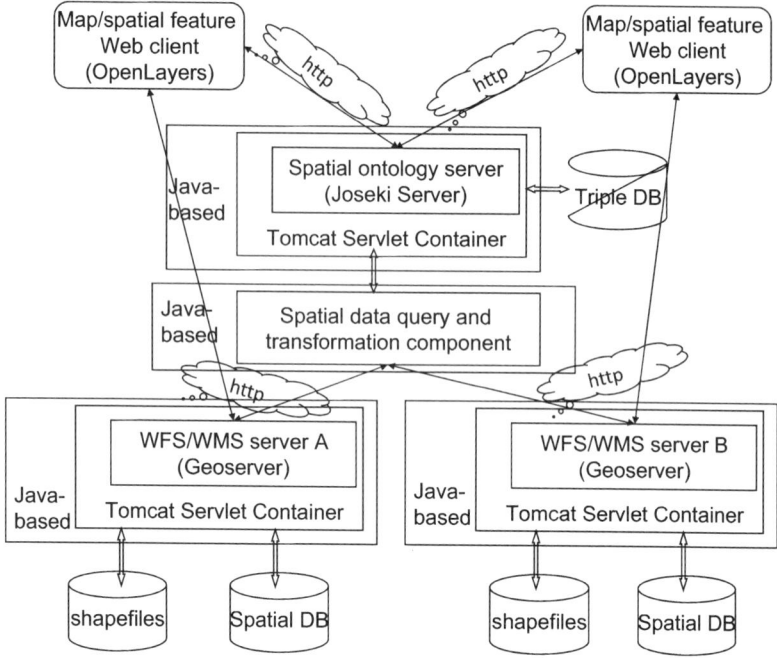

Fig. 6.8 Architecture of the geospatial interface for transit network

features and an application ontology for transportation network data. It stores ontology instances in files or databases. The prototype also includes a spatial data query
and transformation component, which was developed based on Jena library. This
component extracts information from WFS servers and creates ontology definitions
for the extracted web features. Also, the component transforms feature instances
into ontology format to be stored in the ontology server. The last component of the
prototype is a web-based spatial query client based on OpenLayers to render the
ontology queried results as graphic maps.

The data used in the implemented prototype come from the Waukesha Transit Trip
Planning Project. Two WFS servers were created. The bus route WFS server publishes the bus route data using the feature name "wks:routes", while the bus stop WFS
server publishes the bus stop data using the feature name "wksha:BusStops". The
namespace prefixes are different since the features belong to different WFS servers.

6.4.4   Web Service Preprocessing

The prototype implementation includes the following steps for preprocessing data
from web services.

(1) Create a domain ontology for feature data in general and an application ontology for transit bus route and bus stop spatial data. We created three domain specific
ontologies used in transportation networks:

1. transportation base,
2. road networks, and
3. transit networks.

The ontologies are in OWL and translated from the existing UML data models for transportation applications (http://www.fgdc.gov/standards/standards_publications/; Part 7, Transportation Base). We developed algorithms to automatically transform these UML models into OWL ontologies (Zhang et al. 2008). However, because of the differences between UML and OWL, we could not create all necessary ontologies by the automatic transformation method. So we used the ontology editor tool Protégé to create those ontologies that cannot be transformed from existing UML data models. Then we integrated the three domain specific ontologies together. Since the three domain specific ontologies are internally consistent, we successfully avoided all the integration difficulties that would arise from importing other transportation ontologies into ours.

The application ontology in our prototype is divided into two parts: one part corresponds to some WFS features and the other part corresponds to some database tables that supplement the WFS features. To represent geospatial features as ontology instances, we used some predefined OWL classes, such as *Feature, Geometry* (with subclasses *Point, Line,* and *Polygon*), *BoundingBox,* and *Area.* Relations between classes were asserted with the *owl:subClassOf* property. We also introduced some object properties such as *has_geometry* and datatype properties such as *minx, maxx, miny, maxy.* The domains and ranges of properties were asserted using *rdfs:domain* and *rdfs:range.* Restrictions on properties were not used because they might prevent future extension. We further automatically generated some OWL classes from WFS features such as *Route, Link,* and *Stop.* In addition, some OWL properties were auto-generated from feature properties. The scope of OWL names is global, and name conflicts are not tolerated. Because WFS property names are locally scoped, we had to define an OWL property to overwrite any previous definitions of the same names to resolve name clashes. To prevent any further conflicts, we did not place domain or range restrictions on these properties. We related auto-generated OWL definitions with predefined ones through assertions such as *owl:subClassOf* and *owl:subPropertiesOf.* WFS feature instances were automatically translated into OWL individuals using the predefined and auto-generated classes and properties. To enable WFS feature search and discovery, we auto-generated an ontology individual for each feature type to include properties such as feature name, URL, bounding box, and geometry type.

The application ontology corresponding to database tables is a virtual graph of RDF nodes created by a tool—the D2R server from database tables. The RDF ontology was generated based on a mapping configuration file used by the D2R server. The ontology is a straightforward mapping from database tables—one table maps to one ontology class and one column maps to one ontology property. Additional object properties were defined via inference rules based on existing datatype properties and classes. Moreover, we defined inference rules to describe some object properties so as to connect the two parts of the application ontology for supporting the user query. For example, to find the geometry of the bus stops in a bus route, we need both the application ontology that corresponds to WFS features, which contains the geometries and the IDs of the stops, and the application ontology that corresponds to the database tables, which contains the correspondences between a route ID and the IDs of the stops in the route. Using the inference rules and the reasoning ability of OWL ontology, our prototype can support queries that cannot be answered by a WFS or database alone.

(2) Index the available WFS features by extracting feature names, feature property lists, geometry types, and bounding boxes of the features. The indices are instances of a special ontology class*Feature*. Each instance contains the detailed information about a WFS feature. When a user searches for features, the returned feature instances are used to locate the corresponding ontology classes, which are further used to locate all the feature instances.

(3) Map WFS features, properties, and geometries to ontology classes and properties. This process has to be done manually so that ontology classes generated from WFS features will become subclasses of the domain and application ontology classes. Also, if a generated ontology property is equivalent to an existing property, then they are merged. For example, after we generate an ontology class *TransitRoute* corresponding to the feature *wks:routes*, we have to manually identify that this class is indeed a subclass of *TransitLink*, which describes a segment of a transit route. This information cannot be determined based on the names of the features alone since names can be misleading. Similarly, when we generate an ontology property *the_geom* from a common feature property with the same name, if a similar property *geom* has already been defined in our domain ontology to refer to the geometry of features, we need to merge the two properties into one. After completing this process, we can automatically query for spatial data. For example, if we are to find out the geometry of a route by the name of "Summit", we can simply query for instances of *TransitRoute* with the name "Summit". The geometry of the route is the union of the geometries of the links on the route.

(4) Take service queries and return a list of WFS feature services. For example, to locate the requested bus route and bus stop features from the two separated WFS servers, two service queries are needed:

(TransitRoute, { geometry, description}, Line, B), and

(TransitStop,   { geometry, intersection}, Point, B),

where B is a bounding box described in the N3 notation, such as

[a :BoundingBox ;

:maxx   "-73.90782"^^xsd:float ;

:maxy   "40.882076"^^xsd:float ;

:minx "-74.04719"^^xsd:float ;

:miny "40.67965"^^xsd:float

]

These queries ask for features of the type *TransitRoute* or *TransitStop*. The geometry type of the requested feature must be line or point, and its bounding box must
cover *B*. Once a user supplies the two queries, the system matches the ontology
class *TransitRoute* to *wks:routes,* and matches *TransitStop* to *wksha:BusStops*.
Similarly, the property lists are matched against the properties of the two features.
Finally, the system makes sure that geometry types are matched and the two feature services cover corresponding bounding box *B*.

(5) To improve performance, retrieved feature instances are transformed into
ontology individuals and are stored in the ontology server. By this way, clients do
not have to repeatedly send requests to WFS servers for the same feature instances.

6.5 Spatial Decision Support System

6.5.1 Motivation

The Lunan Stone Forest, or Shilin, is the World's premier pinnacle karst landscape.
Located among the plateau karstlands of Yunnan Province in southwest China, it
is widely recognized as the type example of pinnacle karst, demonstrating greater
evolutionary complexity and containing a wider array of karren morphologies than
any other example (Zhang et al. 2005). The area is designated as a national park
covering a protected Shilin area of 350 km^2, and is organized into three zones with
different protection levels. But no much evaluation work was done when the protected-area boundaries were delimited in 1984. The designation of these boundaries
are mainly based on the scenery beautiful values of the Stone Forest Landscape, and
it has no relationship with the karst landscape itself or its natural values. Further the
boundaries are drawn on a small scale (1:1,000,000) geological map. They almost
have no relationship with the topography characteristics such as road, river, topography line, or geological character. Thus, to a great extent it is even difficult for the

administrative officials to know the direction of the boundaries and to find out their exact location, not to say for the public and local residents. This brings difficulty to carry out the conservation regulations.

A web-based SDSS for Lunan Stone Forest Conservation was developed to provide a way to establish rational protective boundaries based on a variety of environmental and social criteria and render the location of the boundaries clear to the public (Zhang et al. 2005). While the web-based SDSS has benefit in many ways from the web technologies such as platform-independent, remote, and distributed computation, the developed web-based SDSS was based on traditional Client-Server architecture and was implemented using traditional computer technologies such as Visual Basic, ESRI Map Objects, and ASP (Active Server Pages) (Zhang et al. 2005). Thus, it is not interoperable and has limitations for share and reuse of geographical data and geoprocessing, although it indeed increased the public access to information and involvement in the decision-making processes for protective boundary designation. The objective of this case study is to develop an interoperable SDSS prototype to assist in protective boundary delimitation for Lunan Stone Forest Conservation. The interoperable SDSS prototype should facilitate share and reuse of heterogeneous geographical data and geoprocessing over the web, and thus can increase the range and depth of information access and improve the solving of decision problems and the effectiveness of decision-making performance. The prototype covers several components in the proposed framework, such as using OGC WFS and WMS services to access the heterogeneous spatial data connected to legacy GIS, using OGC WPS to access the multiple criteria decision model for delimitation of the protected-area boundaries, using ontologies to describe the geographic information, using ontology-based semantic catalogue services to register and discover the published WFS, WMS and WPS services, using the partition refinement algorithm to map local and server ontologies, and using the proposed discovery and composition algorithm to combine web services.

6.5.2 The Prototype

The interface of the implemented interoperable SDSS for Stone Forest conservation is shown in Fig. 6.9. Through the prototype, decision-makers can delimit protective boundaries using the multiple criteria decision model based on a variety of biophysical and social economic criteria by employing ontology-based WFS and WMS. Figure 6.10 shows one scenario of two different level protection areas that were delimited by using the prototype system. Note: in the prototype, the criteria data are stored in two different formats (Shapfile and PostGIS) on two remote servers. The multiple criteria decision model and the ontology-based catalogue software are held in the remote server.

Using the implemented prototype, decision-makers also can render the locations of the boundaries clear to the public by aligning them with conspicuous landscape features such as water bodies, roads or buildings via employing WFS and WMS.

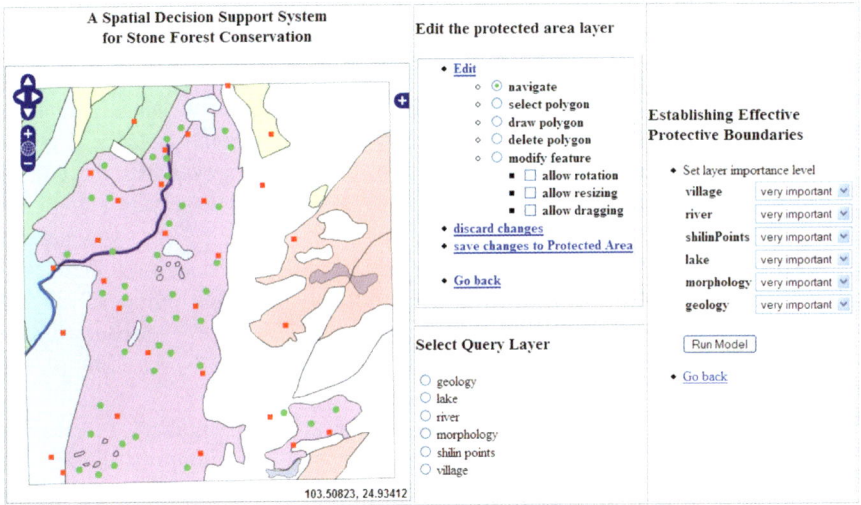

Fig. 6.9 The interface of the web service based SDSS for Stone Forest conservation

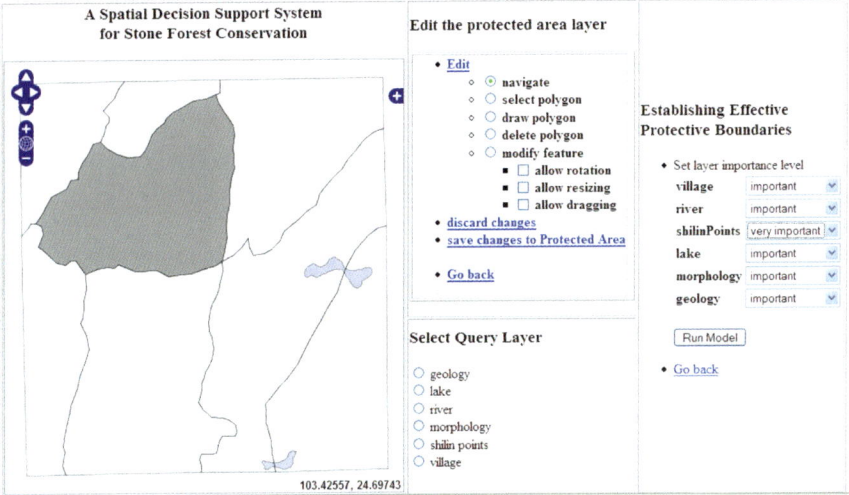

Fig. 6.10 One scenario of two different level protection areas delimited using the prototype system

The following experimental results demonstrate some advantages of the web services-based prototype SDSS:

1. The prototype system provides decision-makers with the ability to access and analyze heterogeneous criteria data in order to make better decisions for protected-area delimitation. It allows the decision-makers access the heterogeneous criteria data from a variety of sources on the Internet. The criteria data, such as

geology, geomorphology, and land use data, are stored in different databases with different formats on separate computers. However, decision-makers can directly access these heterogeneous data sources without having to know specifically who might provide the data and what the format of the data is. Using the prototype, decision-makers can seamless and dynamically integrate the geology data (originally in Shapefile format) located at one data server with the village data (originally in PostGIS databases) located at another data server by invoking the ontology-based WFS and WMS services with little or no knowledge about the heterogeneous environments of the data providers. Through seamless data integration, the web services-based system not only promotes remote access and potential collaborative decision support applications, but also can reduce development and maintenance costs and minimize data conflicts by avoiding redundant data.

2. The prototype system provides decision-makers with the ability of accessing and integrating geospatial data services at semantic level from distributed data sources. For example, the *Shilin point* spatial data used by the SDSS client is composed from two separate WFS web services: in one WFS web service the *Shilin point* feature is named *Yunnan-Shilin*, and in another WFS web service it is called *StoneForest*. Local ontology class *Yunnan:Shilin* has been created to *Yunnan-Shilin* for the first WFS web service, and ontology class *Yunnan:StoneForest* has been developed to *StoneForest* for the second WFS web service. To resolve the heterogeneous local ontology integration problem, the ontology server is used to matching the *Yunnan:Shilin* ontology to the *Yunnan:StoneForest* ontology by using the developed Partition-Refinement algorithm. After the ontology matching, the two semantically heterogeneous WFS web services are composed together into one *Shilin point* spatial data, which is required by the multiple-criteria decision model as one input parameter.

3. The prototype system allows decision-makers access the multiple-criteria decision model across the web via ontology-based WPS. The ontology-based WPS dynamically conducts spatial data analysis, computes the evaluation value and passes the evaluation results as input to ontology-based WFS. The input and output data for the implemented WPS are in GML format, which are connected with WFS. In one scenario of the different level protected-area boundaries calculated by the WPS (shown in Fig. 6.10), the first protection level (dark) covers almost all the limestone pinnacles and the lakes, which are considered by the karst scientists to be of great importance to the landscape; the second protection level (grey) contains less important protection targets, including villages, farmlands, and tourism facilities, such as hotels, commercial stores, roads, and parking lots. Since the multiple-criteria decision model is employed as web services, it provides the interoperable capability of cross-platform and cross-language and can be accessed and reused by other applications and organizations.

4. The web services-based prototype system facilitates decision-makers to access the most up-to-date criteria data. With the ontology-based WFS and WMS, data maintenance of the prototype system becomes easy. Because the criteria data reside in the original databases, they are always updated. Unlike traditional

SDSSs in which the data updated from one source have to be delivered or down-loaded manually to its applications to maintain the changed data, the web services based prototype system automatically propagates the change or update of data. In addition, the web services-based prototype system also allows developers or decision-makers to change or update criteria data or alternative solution maps remotely in disparate sources cross the web. They can create, delete and update geographic features in a remote database over the web using ontology-based WFS. Changes to the protective boundaries are instantaneously relayed to other decision-makers and applications. This instant access to the most up-to-date information enables decision-makers to avoid the tedious process of transferring data and facilitates the decision-making process. In this way, incon-sistencies generated by updates are minimized and enterprise collaboration for a specific joint project is supported.

Besides the aforementioned advantages, the web services-based prototype system has basic GIS functions for enabling data investigation. For example, decision-mak-ers can display and overlay different data layers, and can zoom in, zoom out, pan or query the attribute tables of these data layers.

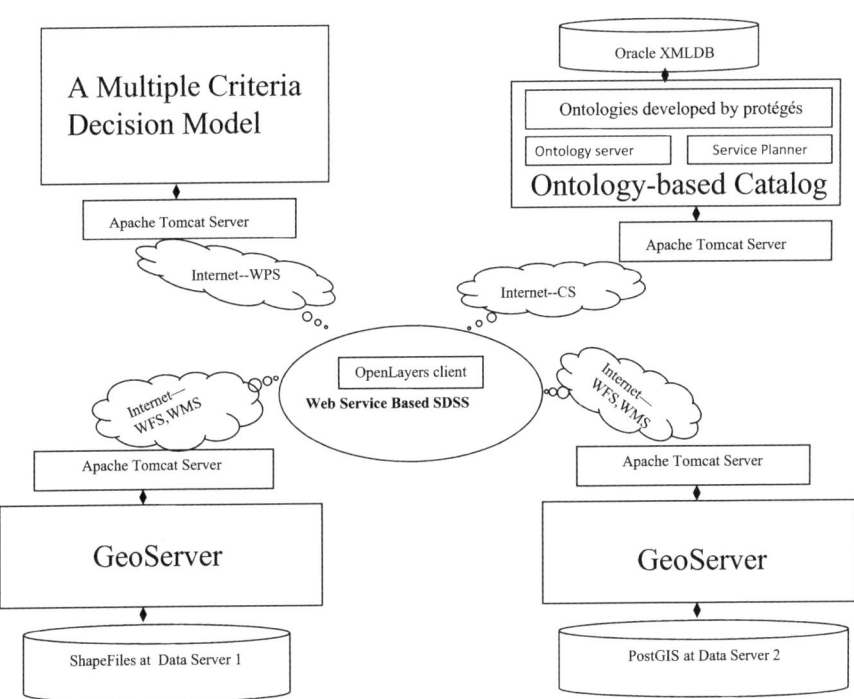

Fig. 6.11 Architecture of the SDSS prototype

6.5.3 Implementation

Figure 6.11 illustrates the prototype architecture, which consists of:

1. Data Service providers: GeoServer instances running in Apache Tomcat to provide WFS and WMS services using ShapeFile and PostGIS backend databases;
2. Web processing service provider: A Multiple Criteria Decision Model to incorporate the interacting biophysical and social-economic criteria, such as geology and geomorphology, into the delimitation of the protected-area boundaries, which is implemented as WPS using Java.
3. Service brokers: An ontology server, which maintains consistency for different local ontologies and is developed using the Protégé editor (http://protege. stanford.edu/), an ontology-based semantic catalogue and service planner
4. Service clients: We implemented the client using the OpenLayers library to provide a user-friendly interface for decision-makers to query and access ontology-based web services such as WFS and WMS. OpenLayers is used for rendering maps and features from WFS/WMS servers and other map services from providers such as Google Map. The service client provides functionalities of querying features from ontology-based WFS servers using OGC filters, retrieving the properties of features selected by mouse actions, and rendering ontology-based WMS maps based on customized style files. Moreover, the implemented service client software supports transactions on ontology-based WFS features that include addition, modification, and removal of features. In contrast to other WFS/WMS client software such as MapBuilder, the implemented client software provides better performance in map rendering and flexibility in editing geometries such as polygon.

The same multiple-criteria decision model applied in the previous web-based SDSS (Zhang et al. 2005) was employed in this prototype, though it is now implemented in Java as web processing services. The multi-criteria evaluation approach was widely used in GIS literature (e.g. Carver 1991). Among many ways to integrate decision criteria, the weighted linear combination method is a popular one (Malczewski 1996; 2000) and was used to delimit different protected-area boundaries in this study. To rank the different protection level alternatives, the following formula was used:

$$S = \sum_{i=1}^{n} W_i C_i$$

where S is the suitability score with respect to the protection objective, W_i is the weight of the criterion i, C_i is the criterion score of i, and n is the number of criteria. The model has its own algorithm to make sure that $\sum W_i = 1$. The weights in the Multiple Criteria Decision Model are classified into three ranks (very important, important, not important) for the simplicity reason and each rank has a default weight. The default weights are obtained based on a survey from domain experts

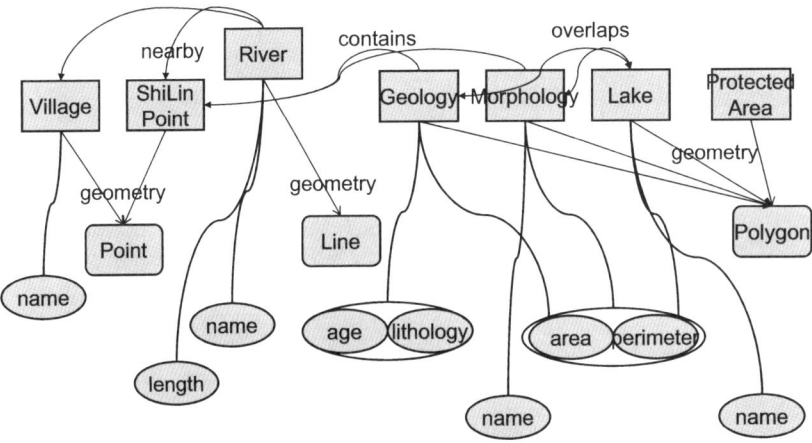

Fig. 6.12 Stone forest ontology: *squares* represent ontology classes, *arrows* between classes are object properties, and *ovals* represent datatype properties

and they indicate the relative importance of the set of criteria based on the preferences of decision makers. Using the weighted formula, overall protective suitability scores were determined and the whole area was divided into several different level protection zones.

6.5.4   Ontology Alignment

The prototype's ontologies describe the features of the stone forest region. Figure 6.12 illustrates part of these ontologies, where *Village* and *ShiLin_Point* are point features, *River* is a line feature, and *Geology*, *Morphology*, *Lake*, and *Protected_Area* are polygon features. Most of the object properties are spatial relations, such as *nearby, contains*, and *overlaps*. One object property—*geometry* relates spatial features to their geometries. There are a few datatype properties as well. The ranges of properties *name*, *age*, and *lithology* are string, and the ranges of *length*, *area*, and *perimeter* are float.

There are many ways to define the local ontologies for our prototype. Thus, it is important that we can align similar local ontologies to an ontology server automatically. To illustrate how this is done, let's consider a variation of the Stone Forest local ontologies shown in Fig. 6.13.

The differences between the two ontology definitions are mostly nominal—the class and property names are different. The object properties are also different by *names* except for *within*. Before we align the two ontology definitions, we reverse the domain and range of *within* and rename it to *within-1*. The ontology classes for representing point, line, and polygon are assumed to be the same. We can apply the partition refinement algorithm (Zhang et al. 2010c) to the two ontology definitions as shown in Fig. 6.14.

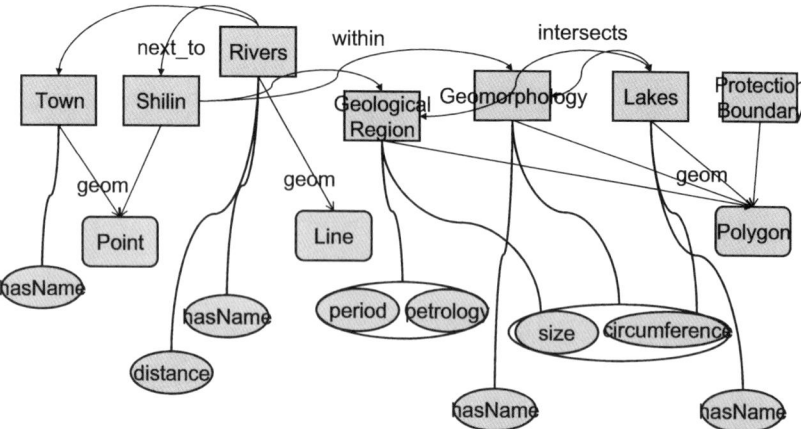

Fig. 6.13 A variation of the Stone Forest Ontologies

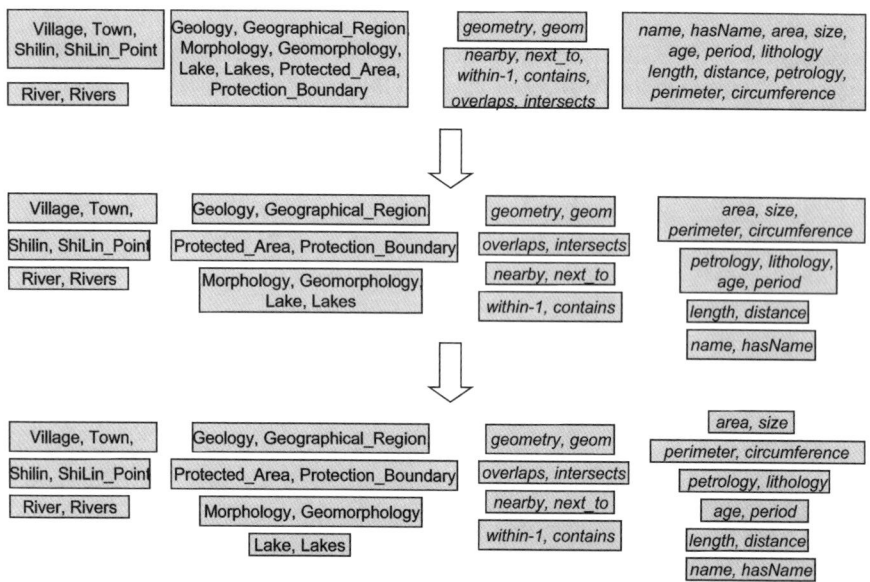

Fig. 6.14 The application example of the partition refinement algorithm to the Stone Forest. Each box represents a partition of classes or properties. The first group is the set of initial partitions. The second and the third groups are partitions after the 1st and the 2nd iterations of the partition refinement algorithm

In the initial partitions, the ontology classes are separate from object properties and datatype properties. Ontology classes that refer to different types of geometries are also separate in different partitions. After the first iteration of the algorithm, class partitions are split according to the numbers of the properties that they have. For example, the set of *Village, Town, Shilin,* and *ShiLin_Point* is split into two partitions, because *Village* and *Town* both have one more property than *Shilin* and *ShiLin_Point* have. The sets of properties are split into smaller partitions because the domains or ranges of these properties are in different partitions. For example, *nearby* and *next_to* are separated from *intersect* and *overlap* since the former group has domains of River and Rivers (line features) while the latter has the domains that are polygon features. Also, datatype properties *name* and *hasName* are split from the rest because their ranges are string and they do not share domains with other datatype properties. After the second iteration of the algorithm, *Morphology* and *Geomorphology* are separated from *Lake* and *Lakes*, since *Morphology* and *Geomorphology* have the properties *contain* and *within-1*, which are no longer in the same partition as *intersect* and *overlap*. In the last iteration, the properties *age* and *period* have to be separated from *lithology* and *petrology* by some other methods, because we cannot tell them apart by domain or range information. We also cannot rely on string comparison since they are very different. In this case, a domain ontology for geographic concepts should be used to find the related terms.

6.6 Chapter Summary

In this chapter, we reviewed several applications of geospatial semantic web that we developed, which include a natural language interface for geospatial data query, a volunteered geospatial information system, a geospatial interface for transit network, and a spatial decision support system. The applications demonstrate the effectiveness of geospatial semantic web in integrating data with heterogeneous semantics. Ontology constructs such as classes and properties can be used to map the definitions of existing data sets to the domain and application ontologies in order to bridge the semantic gap between data sets. After being mapped to ontology definitions, geospatial data can be queried through ontology query language SPARQL, even if the data were originally stored in formats such as ShapeFile, and the databases are accessible through OGC web services. The geospatial data can be transformed to ontology instances that are accessible through ontology services. Alternatively, the geospatial data can be stored in OGC web services while SPARQL queries can be translated to web service requests to query the data. The ontology representation also makes it possible to define a natural language query interface so that text queries can be translated to SPARQL queries to provide a more intuitive query interface.

References

Carver SJ (1991) Integrating multi-criteria evaluation with geographic information systems. Int J Geogr Inf Syst 5:321–339

Dellis E, Paliouras G (2005) Management of large spatial ontology bases. In: Collard M (ed) Proceedings of the first and second VLDB conference on ontologies-based databases and information systems (ODBIS'05/06), Springer-Verlag, Heidelberg, pp 102–118

Malczewski J (1996) A GIS-based approach to multiple criteria group decision-making. Inter J Geogr Inf Syst 10:955–971.

Malczewski J (2000) On the use of weighted linear combination method in GIS: common and best practice approaches. Trans GIS 4:5–22

Peng ZR, Zhang C (2004) The roles of geography markup language, scalable vector graphics, and web feature service specifications in the development of internet geographic information systems. J Geogr Syst 6:95–116

Zhang C, Li W, Day M (2005) Towards establishing effective protective boundaries for the Lunan Stone Forest using an online spatial decision support system. Acta Carsologica 34:178–193

Zhang C, Zhao T, Li W (2010a) Automatic search of geospatial features for disaster and emergency management. Int J Appl Earth Obs Geoinf 12:409–418

Zhang C, Zhao T, Li W et al (2010b) Towards logic-based geospatial feature discovery and integration using web feature service and geospatial semantic web. Int J Geogr Inf Sci 24:903–923

Zhang C, Zhao T, Li W (2010c) A framework for geospatial semantic web based spatial decision support system. Int J Digital Earth 3:111–134

Zhang C, Zhao T, Li W (2013) Towards improving query performance of web feature services (WFS) for disaster response. ISPRS Int J Geo Inf 2:67–81

Zhang C, Zhao T, Li W (2014) Towards an interoperable online volunteered geographic information system for disaster response. J Spat Sci. doi:10.1080/14498596.2015.972996

Zhao T, Zhang C, Wei M et al (2008) Ontology-based geospatial data query and integration. Lect Notes Comput Sci: Geogr Inf Sci 5266:370–392

Zhao T, Zhang C, Anselin L et al (2014) A parallel approach for improving Geo-SPARQL query performance. Int J Digital Earth. doi:10.1080/17538947.2014.904012

Chapter 7
Current and Future Challenges of Geospatial Semantic Web

Geospatial Semantic Web allows users to share geospatial data at the semantic level from multiple semantic heterogeneous sources. However, since the Geospatial Semantic Web is still at the initial stage of research, there are many challenges in implementing a workable system to query, retrieve, integrate, and visualize heterogeneous geospatial data at the semantic level. This chapter introduces several challenging issues faced by the geospatial community to implement the Geopsatial Semantic Web for different applications. These challenging issues include issues in service composition, natural language processing, ontology, performance, big data, and cloud computing. The following sections provide some important examples of current challenges and future research directions for resolving these issues related to Geospatial Semantic Web.

7.1 Service Composition issue

How to develop a service plan to invoke the execution among geospatial web services in correct order needs further research. Many applications such as emergency response and disaster management have complex tasks. Usually an atomic web service may not be sufficient to precisely fulfill a user query. Thus, two or more web services may be needed to synthesize the required services (Lutz 2007). This calls for geospatial web service composition. Geospatial web service composition is the process of reusing and logically recombining existing geospatial web services into composite services to fulfill a user query by providing new functionalities that existing geospatial web services cannot provide alone.

There are two types of geospatial web services composition—syntactic composition and semantic composition. As indicated by their names, the syntactic composition is based on syntactic descriptions while the semantic composition is based on semantic descriptions. For geospatial semantic web service composition, ontologies can be used to describe semantic descriptions.

© Springer International Publishing Switzerland 2015
C. Zhang et al., *Geospatial Semantic Web*, DOI 10.1007/978-3-319-17801-1_7

At present, geospatial web services can be composed using three ways: manual composition, automatic composition, and semi-automatic composition. Syntactically interoperable geospatial web services can be composed manually, while semantically interoperable geospatial web services can be composed automatically or semi-automatically. Initially, geospatial web services were manually composed. Most of syntactic compositions have been done using the manual composition method. In the manual approach, a domain expert usually takes into account user requirements and browses through the available geospatial web services to eventually create a desired composite service. In literature, Web Services Business Process Execution Language (WS-BPEL) (http://download.boulder.ibm.com/ibmdl/pub/software/dw/specs/ws-bpel/ws-bpel.pdf) was developed for dealing with the manual web service composition. In the manual approach, the composite service structure and the atomic services taking part in the composition are described statically, and usually the geospatial web service composition is done syntactically based on workflow. The disadvantage of this manual approach is that users have to intervene in the composition at design time and it is time consuming. With geospatial web services being created and updated on the fly over the Web, it is beyond our ability to analyze them and generate the composition plan manually. In addition, with increasing number of geospatial web services, response time of this manual approach to requests increases dramatically; thus it is unrealistic to expect rapid response using this manual approach.

To overcome the limitations of manual composition and resolve the geospatial web service composition without the need of user intervention, nowadays more and more researchers are beginning to propose different approaches to compose geospatial web services using automatic or semi-automatic ways (Yue et al. 2007; Cruz et al. 2012). The automatic or semi-automatic methods need supports from semantically interoperable geospatial web services. In order to compose geospatial web services automatically, the geospatial semantics of web services must be understandable. Geospatial semantics convey content information about geospatial data, entities, phenomena, functionalities, relationships, processes, services, among others. Geospatial semantics can be classified into four categories: 1) data semantics, 2) functional semantics, 3) execution semantics, and 4) quality of service (QoS) semantics. Data semantics describe the semantics of input and output data in a geospatial web service operation. Functional semantics annotate the semantics for a geospatial web service function. Execution semantics interpret the pre-conditions or post-conditions of a geospatial web service operation. QoS semantics refer to the service selection quality criteria. Ontologies such as the OWL Web Ontology Language for Services (OWL-S) (W3C 2004) can describe geospatial semantics based on service descriptions of the input and output parameters, execution preconditions and post-conditions, and functions. In order to meet the demands of more complex applications and changing environments, more control constructs need to be applied for service composition. OWL-S provides eight kinds of control constructs for service composition, which are: Sequence, Unordered, Choice, If-Then-Else, Iterate, Repeat-While, Repeat-Until, Split and Split & Join. Recently approaches based on OWL-S have been proposed for automated service composition (Yue et al. 2007;

Zhang et al. 2010a). In these approaches, atomic services are described in terms of service ontology and are combined based on a set of control and data flow dependencies among the included services.

Despite the fact that progress has been made for semi-automatic or automatic geospatial web service composition, there are still many challenges for semi-automatic or automatic geospatial web service composition in a distributed and dynamic Internet environment. Currently the automatic or semi-automatic methods can compose web services dynamically by adopting Artificial Intelligence (AI) planning techniques. Most of the existing methods for automated or semi-automated service composition use a sequential structure to compose component services together in the one-dimensional linear fashion. These approaches call web services in accordance with the established order, no matter how diverse the application needs are and how the environment changes. Therefore, they are applicable to only simple cases with a deterministic environment and can neither meet the diverse needs of users nor adapt to the dynamic changes of the environment.

The other problems in semi-automatic or automatic geospatial web service composition are caused by ignoring Quality of Service (QoS), increased response time to request, and reduced performance with increasing number of geospatial web services. One challenge issue in geospatial web service composition is how to obtain best effective services with the composition of services based on maximum quality of QoS and satisfy a user's requirements. QoS defines the non-functional requirements of a geospatial web service such as cost, response time, reliability, availability, accessibility, performance, security, reputation, and integrity.

The "cost" quality is the amount that a service requester needs to pay for executing a geospatial web service. The "time" quality measures the execution time between the service requests sent and results received. It is the length of time for a geospatial web service to provide a response to various types of requests. The "reliability" quality quantifies the ability of a geospatial web service to function correctly and consistently. It is commonly measured using transaction failures per month or per year. The "availability" quality indicates the presence of a geospatial web service to be connected to for a user. It represents the probability that a service is available. It may be associated with time-to-repair (TTR), which is the time it takes to repair a failed geospatial web service. The smaller values of TTR are desirable to achieve high availability. The "accessibility" is the capability to satisfy a geospatial web service request. It represents the achievable speed of a service in time. A great extent of accessibility means that the geospatial web service is available for a large number of clients. The "performance" quality can be used to measure the throughput and latency. Throughput is the number of user service requests served in a given period of time. Latency is the length of time between transferring a request and getting a reply. Larger throughput and smaller latency indicate a good performance. The "security" quality includes authentication, authorization, confidentiality, access control and message integrity. It is an important characteristic of a geospatial semantic web service over the Web. The "reputation" quality measures trustworthiness of a geospatial web service based on the user experience in using the service. It can be calculated by the ranks given by the users for the service. The "integrity" quality is the measure of a geospatial web service in terms of maintaining reliable and correct interactions for a service.

Although the above QoS plays important roles in obtaining best effective composition services, there are only a few studies in literature to investigate automated geospatial web service composition based on QoS requirement (Cruz et al. 2012). Many challenge issues regarding to automated geospatial web service composition based on QoS are still waiting to be resolved in the future.

The ability to perform geospatial web service discovery and composition automatically and dynamically is very important for many geospatial applications. Automated web service composition enables faster responses to user queries compared to the manual case. It also produces up-to-date composition with the latest geospatial web service definitions over the dynamic web environment. However, the dynamic nature of the availability of web services, the large number of alternative combinations of service choices, and real-time requirements on service composition make dynamic service composition a formidable task. Thus, how to create a work flow of services by splitting and joining the available web service choices is still a research topic. It is still challenging to properly compose geospatial web services without a complete and predefined composition plan. It is impossible to define the service composition plan in advance for many applications such as disaster management. In these applications, the service composition plan can only be partially defined. How to compose geospatial web services with a partial composition plan raises a particular challenge. It needs further studies to construct an optimal composite service in a reasonable time without a predefined composition plan and satisfying various requirements. More studies are needed to compose geospatial web services by simultaneously selecting atomic services and inferring the composition patterns among the selected services so as to ensure services are composed in the best way with regard to QoS aggregation.

7.2 Natural Language Processing (NLP) issue

The Geospatial Semantic Web strives to assign proper meaning to the information on the current World Wide Web, so that both human users and machines can use and manipulate the information to better suit their needs. The GeoSPARQL semantic queries on Geospatial Semantic Web can assign proper meaning to the geographic information on the current World Wide Web and overcome the problems that are met in the traditional text/keyword queries. For example, the traditional text/keyword queries require an exact word from the query to appear in the searched GML data/metadata. If a mistake is made or a word is used in a different form/name (synonyms/hyponyms) than in the data/metadata, users may not find the right answer. Through assigning proper meaning to the information on the Web, the GeoSPARQL semantic queries overcome these problems and can improve the query processing capabilities of GML data based on semantics derived from ontologies. The GeoSPARQL semantic queries attempt to obtain geospatial features without knowing their detailed syntactic structure. Unlike syntactic queries such as XPath and XQuery queries for GML, which only support retrieval of explicit data based on syntactic information, the GeoSPARQL semantic queries enable retrieval of both explicitly and implicitly derived information based on syntactic and semantic information contained in the spatial data.

Despite these advantages, the GeoSPARQL semantic queries may present a challenge to the general users who have no training on how to use GeoSPARQL or the spatial data set that they wish to query. Although GeoSPARQL is a standard way to access RDF data, it is tedious and difficult for end users because of the complexity of the GeoSPARQL syntax and the RDF schema. While experts can easily gather information from the wealth of spatial data from the Geospatial Semantic Web by using GeoSPARQL, most lay users lack the expertise necessary to proficiently interact with the applications in the Geospatial Semantic Web. Consequently, non-expert users usually have to rely on forms, query builders, question answering or keyword search tools to access RDF data. However, these tools have so far been unable to explicate the queries they generate to lay users, making it difficult for these users to assess the correctness of the queries generated out of their input and to check whether the retrieved answers indeed correspond to the intended information. An ideal Geospatial Semantic Web should allow end users to profit from the expressive power of the Semantic Web standards such as RDF and GeoSPARQL while at the same time hiding their complexity behind an intuitive and easy-to-use interface. A natural language processing (NLP) interface may help to achieve this goal. To help end-users acquiring the needed geospatial information from the knowledge base over the Geospatial Semantic Web, a natural language processing (NLP) interface may be developed to allow them express their needs without having knowledge in ontology or GeoSPARQL.

Compared with the "artificial language" such as GeoSPARQL, a natural language is the language used by the general public for daily communication. An "artificial language" such as GeoSPARQL is characterized by self-created vocabularies, strict grammar, and a limited ideographic range, and therefore is less easy to be accustomed to, and not easy to be understood by the general users. However, the syntactic and semantic flexibility of a natural language enable it to be natural to human beings. The natural language can be easily understood and expressed by the general users.

Natural language processing (NLP) studies how to enable a computer to process and understand the language used by human beings in their daily lives, to comprehend human knowledge. Natural language input may be a convenient way for the end-users to retrieve information from Geospatial Semantic Web, because there is no need to summarize his/her thoughts in keywords. Using the NLP interface users can merely type the needed information in natural language in order to retrieve the desired information.

The NLP interface takes input queries using natural language expressions and sends queries to multiple data sources through the ontology-based knowledge base on the Geospatial Semantic Web. By using the NLP interface, the general users should be able to intuitively find out the requested geospatial information without special GIS or programming training. The NLP interface should take advantage of the semantic markup and ontologies to interpret the natural language queries and expose semantic geospatial web services directly to the general users. Therefore, the NLP interface should be able to dramatically low the barrier of the various application users' entry to the Geospatial Semantic Web. With the NLP interface it becomes unnecessary for the general users to understand formal ontologies and precisely define vocabularies. Thus it may extend the targeted users of Geospatial Semantic Web from those who are familiar with GIS and Geospatial Semantic Web technologies to the general public.

However, there are many challenges in the development of the NLP interface because of the ambiguity of the natural language. It is difficult to transform natural language queries into formal GeoSPARQL queries because it is necessary to correctly map the vocabulary of the natural language to the vocabulary of the knowledge base on the Geospatial Semantic Web.

To simplify mapping of the vocabulary of the natural language to the vocabulary of the knowledge base, only domain specific ontologies are considered to produce a lexicon for translating user input in the existing literature, and application ontologies have been totally ignored (Zhang et al. 2010b). For example, in Zhang et al. (2010b) the lexicon was constructed automatically from the domain ontologies in the ontology server that created the verb, noun and prepositional phrases with the relations in the ontological structure. The lexicon in their proposed framework was composed of three sources: (1) domain ontology entities in the ontology server, including domain ontology classes (concepts), domain ontology properties (relations), and domain ontology instances (individuals), which were used to limit the ambiguities and errors in the natural language interactions; (2) general dictionaries, such as WordNet, which were used to enlarge the vocabulary of the domain ontology and help mapping user vocabulary to domain ontology vocabulary; (3) application specific synonyms, such as user-defined synonymy words, which were used to define application jargons and abbreviations. For the simplification reason, application ontologies have not been considered in the study (Zhang et al. 2010b). However, the vocabulary of the knowledge base on the Geospatial Semantic Web should not be limited to the vocabulary of the domain ontologies. The vocabulary of the knowledge base on the Geospatial Semantic Web should also consider the vocabulary of the application ontologies. However, it is difficult to consider the various application ontologies. In addition, due to its endless exceptions, changes, and indications, a natural language also becomes the type of language that is the most difficult to be mastered. The coverage by natural language is much larger than the coverage by ontologies. The constraints imposed by ontology structure also make it difficult to map the vocabulary of the natural language to the vocabulary of the knowledge base expressed by ontologies on the Geospatial Semantic Web. Therefore, it remains a research challenge to completely map the vocabulary of the natural language to the vocabulary of the knowledge base on the Geospatial Semantic Web.

In addition, ontologies on the Geospatial Semantic Web describe little time dependent information; thus they have limited capability to reason about temporal issues. Therefore, the NLP interface still cannot cope with retrieving temporal geospatial information. It is difficult to extract temporal geospatial information based on simple questions formed with "how long" or "when". It is also difficult to handle questions expressed using time phrases such as "in the last year" because the structure of temporal data is domain dependent. It is still a research challenge in making inferences from temporal expressions and temporal relations and determining how to handle temporal data in a way which would be portable across ontologies.

In addition, because it is difficult to develop a Linguistic component to fully exploit quantifier scoping such as "each", "all", and "some", the current existing systems using the NLP cannot understand queries formed with "how much". However, unlike temporal data, some basic aspects of quantification are domain independent; thus it may be easier to resolve this issue in the near future.

Finally, the spelling or grammatical errors make NLP far from perfect. Automated grammatical error detection and correction have been focused on NLP over the past dozen of years. While grammatical error detection and correction are relatively easy to resolve for NLP, spelling errors are difficult to detect, because some words to be corrected are possible words in English, such as confusion (e.g. form/from), split (e.g. Now a day/Nowadays), and derivation (e.g. badly/bad) errors. The conventional pipeline for grammatical error detection and correction has a limitation due to the different types of spelling errors and the unavailability of contextual information. More research for dealing with not only typographical errors but also spelling errors, such as confusion, split, and derivation errors, is needed.

In general, although natural language facilitates describing contents and can be easily understood by human users, the vocabulary of natural language is ambiguous and very hard to be reasoned about by Geospatial Semantic Web. It is still a challenge to develop NPL across domains because heterogeneous vocabularies may be used to describe similar information across different domains. Thus, even if a user's terminology may be translated into semantically sound triples containing terminology distributed across ontologies, it still has problems. In addition, nowadays NLP is much affected by the big data problem. The heterogeneous nature of a language ranging from a free-styled email communication to semi-structured Wikipedia articles in Big data brings more challenges for NLP technology. To sufficiently address the challenges, the NLP technology should be designed in a highly scalable manner. The goal of the Geospatial Semantic Web is that web content should be expressed not only in a natural language, but also in a language that can be unambiguously understood, interpreted and used by computers and human users, thus permitting them to find, share, and integrate information more easily. It is still a long way to go to achieve this goal.

7.3   Ontology issue

Ontology is used to resolve the semantic heterogeneous problem over Geospatial Semantic Web, and ontology development forms the backbone of Geospatial Semantic Web (Wiegand and García 2007). Therefore ontology quality and ontology matching are important for Geospatial Semantic Web. However, ontology quality and ontology matching remain a challenge for development of Geospatial Semantic Web.

Ontology quality may be the most important challenge for development of Geospatial Semantic Web. Although some studies have created some specific application ontologies, currently these ontologies are typically built by a small number of people, in most cases by researchers, using ontology tools and editors such as Protégé. Although these ontology tools and editors supporting ontological modeling have been improved over the last few years and many functions are available now, such as ontology consistency checking, import of existing ontologies, and visualization of ontologies, ontology building manually has proven to be a very difficult and error-prone task and becomes the bottleneck of knowledge acquiring processes. For instance, it is unrealistic for non-domain-experts to use these tools to build high quality ontologies. Although transformation algorithms have been proposed by Zhang et al. (2008) to automatically transform the existing UML to OWL so as to avoid errors and provide a cost efficient method for the development of high quality ontologies, there are many issues yet to be resolved due to the differences between UML and OWL.

With more onotologies constantly grow in size, it is more difficult to understand, maintain and edit ontologies. The online tools for collaborative ontology development may be a good way for creating high quality ontologies. The online tools for collaborative ontology development are usually more efficient for performing ontology maintaining and editing. However, many existing ontology tools do not have the web-based interface for collaborative ontology development. Several issues need to be considered to facilitate a collaborative ontology development. One issue is how to implement a transaction management to avoid problems when many users create or edit the ontologies and commit their changes simultaneously. While the concurrency control may be easy to solve this problem on the database level, it is challenge to solve this problem on the scalable web level. The second issue is how to implement a locking mechanism on ontology module level to avoid inconsistencies of online ontology creation and edition. When different members of geospatial community collectively extend an ontology, different opinions inevitably can cause conflicts to occur. Consequently, a conflict resolution should be an essential functionality. However, it is not easy to resolve the conflict and guarantee the correct decisions, therefore, it needs further studies as to how to solve this conflict problem and develop a rating system such that performed ontology editing operations can be rated. When a user plans to insert a new element into existing ontology, to avoid duplicated concepts a search engine may be needed to make sure that the concept does not exist in existing ontology. However, it is challenge to implement such a search engine. More studies are needed in this direction.

Ontology matching is another important challenge for development of Geospatial Semantic Web. On Geospatial Semantic Web different users communicate and collaborate based on different ontologies connected with different knowledge expression means. The differences between ontologies from varied sources can be handled by ontology matching. Ontology matching is a solution to the semantic heterogeneity problem over the Web by finding correspondences between semantically related entities of ontologies, and ontology matching is inevitable for ensuring interoperability in Geospatial Semantic Web. Ontology matching is important for various tasks in Geospatial Semantic Web such as query answering and data translation.

There are two major aspects of an ontology to affect the ontology matching process. The first is the complexity of labels used to describe classes, relations and instances in the ontology. This has an influence on the initial determination of candidate correspondences. The second is the complexity of the structures used to define these elements that is often used to improve and validate the initial hypotheses. Different approaches of ontology matching have been proposed in the literature. The main distinction among them is due to the type of knowledge encoded within each ontology, and the way it is used when identifying correspondences between features or structures within the ontologies. For example, *Terminological* methods lexically compare strings used in naming entities, while *Semantic* methods use model-theoretic semantics to decide whether or not a correspondence exists between two entities. The past several years have witnessed impressive progress in the developing of ontology matching tools. However there are many issues waiting to be solved for reliable ontology matching.

The first issue is that ontology matching needs to work on a larger scale. In the era of "Big Data", ontology matching needs fast algorithms to analyze and integrate a large set of ontologies. However, existing ontology matching tools have not demonstrated that they can handle a large set of ontologies.

The second issue is that performance of existing ontology matching tools needs to be improved. For dynamic applications in Geospatial Semantic Web, performance is particularly important because users cannot wait too long for the Web to respond. However, ontology matching may become bottleneck for the dynamic applications if the matching techniques perform slowly.

The third issue is missing background knowledge. Missing background knowledge is one important reason for failure of ontology matching. Ontologies are developed with certain background knowledge and in a certain context. However, the background knowledge and context information may not be available for matching tools. The lack of background knowledge and context information may generate ambiguities thus increase uncertainties and errors of ontology matching. Strategies are needed to resolve the missing background knowledge and context information in ontology matching. Because uncertainty exists in ontology matching, probability based on mapping techniques (with a probability attached to each mapping) may be used for reducing uncertainty iteratively. Probabilistic reasoning may be used to improve detection of mapping inconsistencies.

In addition, the lack of sufficient metadata annotating of the ontologies is also a challenge for the development of Geospatial Semantic Web. While ontologies are important for semantic interoperability and communication among data sources, ontologies are always developed by groups or individuals in isolation. There is lack of metadata annotations of ontologies, which causes difficulty for ontology matching and sharing. The various ontologies without metadata are usually developed using different techniques. There is no any enforced standard convention for describing the contents and context of the ontologies. However, metadata annotating the ontologies should facilitate ontology discovery and match over Geospatial Semantic Web.

Finally, it is challenge to provide a dynamic ontology matching support infrastructure at the web scale, so that tools and applications can rely on it in order to share, publish, and reuse matching tools. The matching life cycle is tightly related to the ontology life cycle: as soon as ontologies evolve, new matching tools have to be produced following the ontology evolution. This may be achieved by recording the changes made to ontologies and transforming these changes into matching processes, which can be used for computing new matching that will update the previous ones.

In general, nowadays web sites are no longer static web pages serving content and images any more. They become more responsive, adaptive, and dynamic. There is inheriting uncertainty in the ontology creation, maintenance, and matching processes. The use of newer data formats (many of which are schemaless) makes it harder to use existing ontology creating, matching and alignment techniques for the current web applications. Under these conditions existing approaches for ontology creation, maintenance, and matching need to be modified and new perspectives for solving this problem need to be developed for the future. Since the existing approaches based on deterministic assumptions will not perform well in situations that are non-deterministic, probabilistic methods based on approximate sampling may be explored to overcome this problem. In dynamic settings of the Web, it is natural that applications are constantly changing. Thus approaches that attempt to tune and adapt automatically ontology creation and matching solutions to the settings in which application operates are of high importance. However, it is hard to perform automatic tuning and adapting. It is too cumbersome for one person or a small group of people to resolve the problem. Many people need work together for creating high quality ontologies and matching correct ontologies. Crowd sourcing, collaborative and social approaches, which allow easily sharing and reusing, may be used to aid in ontology creating and matching.

7.4   Performance issue

Geospatial Semantic Web extends the Web from a data archive and infrastructure to a knowledge engine, which enables more powerful reasoning and information retrieving from heterogeneous and contradicting conceptual models and scientific data in the Web. Geospatial Semantic Web promises better retrieval of geospatial information by explicitly representing the semantics of data through ontologies, which can be understood and processed by computers. It also promotes sharing and reuse of spatial data for a wide variety of applications by using standardized Semantic Web languages such as GML, WFS, RDF, Geo-SPARQL to encode spatial data.

However, representing structured geospatial data in these languages can result in inefficient data access. Literature shows that obtaining spatial information becomes very slow when query the WFS systems from large geospatial databases over the Internet (Peng and Zhang 2004; Zhang and Li 2005). This is because the WFS transport text-based GML data over the network. When GML-coded geospatial data are transported, all the markup elements that describe spatial and non-spatial features, geometry, and spatial reference systems of the data are also transported to the recipient. While GML is important for data interoperability and the GML-coded data could be saved and used by any other client-side applications that can read GML data, GML also greatly slows down the performance for the system. Compared with some binary GIS data formats, the size of GML data files is large. Large file sizes may hinder the use of GML files as a means of data transport over the Internet.

Further, many applications such as disaster response applications require many users concurrently access spatial databases through highly intensive geocomputation processes. The spatial data objects are generally nested and complex, and spatial queries are based not only on the attributes of spatial objects but also on the spatial locations, extents and measurements of spatial objects in a reference geographical system. Therefore, spatial query requires intensive disk I/O accesses and spatial computation. This brings further challenges for efficiently and fast conducting spatial queries from the Geospatial Semantic Web systems. With the development of Spatial Data Infrastructure and GIS, the spatial data are growing exponentially year by year and they are becoming more diverse. While GML and WFS facilitate spatial data sharing and provide feature-level data search, access, and exchange over the Web thus decision makers need not download a whole data file for analysis, performance is becoming an important issue for users concurrent accesses to spatial information, which may be obtained by involving highly intensive computation. This has been acknowledged widely in literature. However, there are a few studies in literature to investigate the ways to improve the performance of GML and WFS (Zhang et al. 2013).

In addition, one of the main obstacles that prevent efficient and distributed query on geospatial knowledge base is the lack of indexing on spatially related data objects. This problem is inherent in the RDF representation of spatial data, which consists of loosely connected data objects related by object properties. Even if spatial objects are originally stored in related database tables, once they are transformed to RDF objects, the spatial indices are lost. It is possible to recreate indices for RDF objects with spatial attributes. However, pre-computing spatial indices does not guarantee performance improvement since the RDF queries are much more flexible than database queries and it is difficult to predict which spatial objects should be indexed and how. Thus, it is necessary to implement suitable extensions of the RDF query engine to take advantage of the creating spatial indexing on-the-fly.

Moreover, the Geo-SPARQL protocol, which was proposed by OGC (Open Geospatial Consortium) as an extension of SPARQL for querying geographic RDF data, is dominated by spatial join operations due to the fine-grained nature of RDF data model. Lack of spatial indices causes additional performance problems for Geo-SPARQL queries. One reason for the poor performance problems is caused by the way that spatial attributes are stored in RDF data sets. Spatial attributes are usually stored as string literals that conform to certain formats such as WKT or

GML. The Geo-SPARQL query engine that implements spatial operators and filter functions has to parse these strings to recover the spatial coordinates for spatial computation. A naïve implementation of a spatial operator or a filter function in Geo-SPARQL treats its spatial inputs as plain strings and has to parse the strings to retrieve spatial contents such as X and Y coordinates. Repeated parsing of the spatial inputs imposes a very large runtime overhead. The second reason for the poor performance problems is due to lack of parallelization. Since spatial objects are not indexed, Geo-SPARQL query engine cannot partition ontology data into subsets to be processed in parallel. As a result, Geo-SPARQL query can only be processed as a single-threaded program. Even with pre-computed spatial indices, partitioning spatial ontology data is not easy since the targeted data may not be evenly distributed in the indices.

Zhao et al. (2014) introduced a new parallel approach for improving the query performance of geospatial ontology in a Geo-SPARQL query by separating spatial and non-spatial components. In fact, different parallel approaches have been widely used for improving query performance for a long time in literature. However, past research on improving query performance using parallelization has been centered on relational databases (e.g. Boral et al. 1990; DeWitt et al. 1986). Optimizing techniques for parallel relational databases do not specialize on the triple model of RDF and triple patterns of SPARQL queries for query engines based on the RDF and SPARQL-specific properties (Groppe and Groppe 2011). Although there are studies to query heterogeneous relational databases using SPARQL and parallel algorithms (e.g. Castagna et al. 2009; Karjalainen and Kemp 2009), parallel relational databases have inherent limitations such as scalability. SPARQL query can be parallelized by treating each triple statement in the query as a parallel task and the results of all the triple statement sub-queries can be joined together after all the parallel tasks have completed (Groppe and Groppe 2011). Unfortunately, this approach does not work efficiently when spatial predicates exist in the triple statements. There are also studies to propose methods for efficiently parallelizing joint query of RDF data using Map-Reduce systems (e.g. Ravindra et al. 2011; Anyanwu 2013). However, except Zhao et al. (2014) there is no study to deal with parallelizing spatial join computations to support efficient spatial RDF query, which is an important issue for the development of Geospatial Semantic Web (Zhang et al. 2007).

To improve the query performance of WFS, we proposed a parallel approach, which makes full use of the Voronoi diagram for creating a spatial index and data/task parallelism for concurrent spatial queries, to execute the query processes on a large cluster of computers (Zhang et al. 2013). The proposed approach is a geography aware method to partition a large map region. It splits and indexes the input data by constructing a Voronoi diagram. A Voronoi diagram is designed to guarantee that data within a partitioned region are stored on one node and all spatial data are distributed across clusters according to geographical space. A Voronoi diagram is extremely efficient in searching a nearest neighbor region because it divides the two dimensional geo-space into several parts, and each part records the nearest relationship through the shared edge between each pair of neighbor parts. The experimental results show that the Voronoi diagram indexing + data/task parallelism processing architecture can reduce individual spatial query execution time by taking advantage of parallel and distributed processes thus can afford a large number of concurrent spatial queries. In the study (Zhang et al. 2013), to obtain the needed information from diverse distributed data sites, a user's query will be first decomposed into a sequence of sub-queries. The decomposed spatial queries are then analyzed and translated into parallel programs for locating data fragments in data sites, query optimization, and query execution. After that, the parallel query results from multiple WFS servers are retrieved and combined by performing spatial calculations. Finally, the combined geospatial features are delivered as a single response to disaster responders or decision-makers. The basic idea of the proposed approach is to locate all distributed data sources that might contain answers to a user query. In a distributed system, data required for query processing need to be located since it may be present across several data sites.

Recently we also proposed a novel approach to optimize and parallelize spatial joins in Geo-SPARQL queries (Zhao et al. 2014). The novel idea of the proposed parallel approach is to separate spatial and non-spatial components in a Geo-SPARQL query. Instead of pre-computing spatial indices for geospatial ontology and implementing spatial extensions of a query engine to use the indices, we proposed a strategy for a query engine implementation by separating spatial from non-spatial components in a Geo-SPARQL query and processing spatial sub-query after non-spatial sub-queries have been completed. The main benefit of this approach is that a smaller set of ontology objects can be obtained after non-spatial sub-queries so that their spatial attributes can be cached for subsequent spatial computation including on-the-fly spatial indexing and spatial joins. Since the parsed spatial attributes are cached, the overhead caused by repeatedly parsing of spatial literal strings can be avoided. The results of this research show that the proposed approach can facilitate the access to spatial ontology information for multiple users through highly intensive geo-computation processes over Geospatial Semantic Web, particularly for time-critical applications such as disaster response.

The parallel approaches (Zhang et al. 2013 and Zhao et al. 2014) can also be combined with other strategies to further improve the query performance of Geospatial Semantic Web queries. For example, with the growth and widespread use of smart phones, in the future many applications may use smart phones to query the needed spatial feature information from the Geospatial Semantic Web for updating and sharing the geospatial information. With the wireless network for smart phones, a zipped binary XML (BXML) or the popular GZIP-based compression algorithm may be employed to decrease the data volume in the network transmission to further improve the efficiency of Geospatial Semantic Web performance.

The proposed parallel approaches (Zhang et al. 2013 and Zhao et al. 2014) can also be combined with the caching strategy to further improve the efficiency of query performance over the Geospatial Semantic Web. The frequently queried results can be cached on the Geospatial Semantic Web servers to allow the systems to respond as quickly as possible. Caching reduces the data access time by storing results of frequent spatial joins in memory so that the same operation is not repeated for every data request.

In addition, the proposed parallel approaches (Zhang et al. 2013 and Zhao et al. 2014) can also be combined with the progressive transmission technique to improve the efficiency of query performance from the Geospatial Semantic Web. With the progressive transmission technique, the spatial features in the Geospatial Semantic Web are extracted using cartographic principles to construct the multi-layer structure. Several methods have been proposed in literature for the vector data progressive transmission (Buttenfield 2002). These methods can be used with the proposed parallel approach to further improve the efficiency of query performance from the Geospatial Semantic Web.

Finally, the proposed parallel approaches (Zhang et al. 2013 and Zhao et al. 2014) can be easily adapted to Cloud computing services to further improve the performance. Cloud computing can provide distributed computing capability in elastic and on demand manners by virtualizing and pooling computing resources. By providing "computing as a service" for end users in a "pay-as-you-go" mode, cloud computing may be more convenient and budget and energy consumption efficient for improving the performance of the Geospatial Semantic Web systems of heavy workload.

However, it remains a challenge to combine the proposed parallel approaches (Zhang et al. 2013 and Zhao et al. 2014) with the aforementioned other methods to further improve performance of the Geospatial Semantic Web queries. There are also other issues that need further studies. For example, how to efficiently handle large geospatial knowledge bases in the ontology-based search engine and how to efficiently handle geospatial data reasoning are also waiting for additional study. A search may become slow for data with a large number of geospatial features. While DL is well suitable for the representation of structured or semi-structured attribute information, it becomes complicated if geospatial data is considered. For example, for knowledge bases it would be necessary to be able to compute spatial relationships from the geometry of objects. However, the DL-based system is unlikely to efficiently handle large geospatial knowledge bases. The current reasoners are less capable when handling large amounts of geospatial data and expressive ontologies, and there remain daunting challenges in building reasoners supporting geospatial data and the full use of ontology languages.

7.5   Geospatial big data issue

"Big Data" is a term to describe the volume of data on the Web which becomes so large and complex that it is difficult to process using the traditional data management and processing tools and it is hard to analyze it to extract relevant meaning for decision making (Barrett et al. 2013; Michael and Miller 2013). "Big Data" include not only structure data but also unstructured information such as text and imagery. The three important characteristics of "Big Data" are high volume, high variety, and high velocity (Laney 2001).

With the development of computer and Internet technologies, the amount of data produced and communicated over the Web is rapidly increasing. Every day around 20 quintillion (10^{18}) bytes of data are produced (http://www-01.ibm.com/

software/data/bigdata). This trend will only accelerate. It is estimated that by 2020 more than 50 billion devices will be connected to the Internet (http://share.cisco.com/internet-of-things.html). Data including geospatial data are produced from many different sources and platforms such as enterprise, social media and sensors. Because of recent development of GPS and sensor devices the cost of geospatial data acquisition has dramatically cut down. Low-cost GPS and sensor devices such as smartphones and wireless sensor nodes support the growth of geospatial data collection and communications. Many sources such as sensors, humans, and applications start generating geospatial data and tend to store them for long time due to inexpensive storage and processing capabilities. The collection of spatial data sets become so large and complex that traditional data management and processing tools are difficult to be applied to the geospatial big data over the Geospatial Semantic Web. It is well-known that the geospatial big data will only be of value if the data can be collected, analyzed, and interpreted. However, it is difficult to capture, store, search, share, transfer, analyze, and visualize the geospatial big data over the Geospatial Semantic Web. It is more difficult to acquire useful information from boundless geospatial big data.

One of the key challenges in making use of geospatial big data lies in finding ways to deal with heterogeneity, diversity, and complexity of the data. Traditional technologies cannot deal with these issues and acquire useful knowledge from the geospatial big data. Semantic technologies are meant to deal with these issues and can make data machine readable and process data intelligently. Geospatial Semantic Web technologies allow finding, sharing, and integrating geospatial big data in a meaningful way. Without the meaningful use of geospatial big data, it is difficult to efficiently manage geospatial big data at unimaginable scale. Therefore, Geospatial Semantic Web technologies may solve some of the problems identified for geospatial big data. Geospatial Semantic Web technologies may make data integration automatic and easier. However, for Geospatial Semantic Web the geospatial big data will not be stored in one or a few locations. They will not be just one or even a few types and formats and will not be amenable to analysis by just one or a few analytics. There will also not be just one or a few cross linkages among different geospatial big data elements. Thus it is challenge to make effective use of geospatial big data on Geospatial Semantic Web. More studies need to be carried out in the future.

The second challenge is efficiently managing geospatial big data at unimaginable scale. Because geospatial big data are often stored at different locations and data volumes may continuously grow, an effective computing platform will have to take distributed large-scale data storage into consideration for computing. The traditional data management technologies that require all data to be loaded into the main memory may become a technical barrier because moving geospatial big data across different locations is expensive, even if we do have a super large main memory to hold all geospatial big data for computing. While communication hardware speeds are increasing with new technologies, message handling speeds are increasing only slowly. In addition, the existing ontology reasoned on Geospatial Semantic Web cannot process mass geospatial big data. For example, Jena and Pellet, the two existing popular ontology reasoners, can only process triples not more than 10 millions in a single computer. It is waiting for further study as to developing a powerful ontology reasoner for geospatial big data on the cluster computers using the Cloud computing technology. Geospatial big data need massively parallel software running on many servers to manage and process ontology reasoning within a reasonable time. Now it is still difficult to efficiently extract useful geospatial knowledge information from the geospatial big data. Moreover, general computational solutions, especially using unstructured geospatial big data, are not known yet.

The third challenge is the quality of geospatial big data. The Web is an open medium in which anyone can publish data. Therefore, the geospatial big data served by the Web contains data that is outdated, conflicting, or intentionally wrong such as SPAM. Uncertain and incomplete data are normal for geospatial big data applications. For example, the geospatial big data generated by GPS are inherently uncertain caused by the precision limitation of the GPS equipment. Although the availability of geospatial data or information has increased, there are missing or incomplete data. For example, failure of a sensor node may miss some data. How to handle missing or incomplete data is one of the important future research topics for geospatial big data. Traditional technologies, which assume semantically homogeneous data, cannot perform meaningful data management and integration of geospatial big data over the Geospatial Semantic Web. In addition, it is challenge to assess the quality of geospatial big data and determine the trusted available data. Given the large volume, it is impossible to validate every data item of geospatial big data over the Geospatial Semantic Web. New approaches to qualify and validate geospatial big data are needed.

The fourth challenge is privacy of geospatial big data. As more data is accumulated about individuals through social media, google mapping, and health information, more people are worrying about violations of one's privacy and they fear that certain organizations will know too much about individuals. Therefore, how to develop methods and algorithms that randomize personal geospatial information among a large data set enough to ensure privacy is a key research problem. In addition, geospatial big data should be secured with respect to privacy and security laws and regulations. More studies are needed to clearly define these security levels and map them against both current law and current analytics. For example, if Twitter runs an analytic over its databases to extract an individual's geospatial information, at what security level should that analytic operate? Data ownership in the social media arena also presents a challenge. Who own the petabytes of social media geospatial big data reside on the servers of Twitter, YouTube, and Facebook? Of course, the social media companies are contending for the ownership because of residency. Undoubtedly, the "owners" of the pages or accounts believe they own the data. Who really owns the geospatial big data on social media? This question is still in the debate stage and data ownership is hard to decide because our laws have focused on physical assets that couldn't be duplicated. How shall society protect itself against those who would misuse or abuse large databases? What new regulatory systems, private-law innovations or social practices should be? Because "geospatial big data" are so novel that many questions need to be answered in the future.

The fifth challenge is lack of tools and trained personnel to properly work with geospatial big data. While it is assumed that geospatial big data will bring more information and impacts across many different fields, the lack of tools and trained personnel raises concerns about the successful use of geospatial big data. Although our analytical capabilities and tools have been improved overtime, they are not designed to deal with geospatial big data. New tools and mechanisms are needed to covert latent, unstructured text, image or audio information into numerical indicators to make them computationally tractable. Otherwise, only a very small percentage of all the geospatial big data available on the Geospatial Semantic Web will be analyzed and most data will go un-analyzed. While most analytics vendors such as Teradata or Vertica claim being able to handle multi-petabyte databases, the database tools are difficult to deal with petabyte-scale collections of geospatial big data that come from click streams, transaction histories, sensors, and elsewhere. Because databases have to slowly import data into a native representation before they can be queried, and this limits their capability to handle streaming geospatial big data. The current existing streaming technologies for databases cannot integrate streaming geospatial big data well into their relational engines.

In summary, with the development of Internet and GPS technologies, the geospatial information and data is unbelievably large in scale, scope, distribution, heterogeneity, and supporting technologies. Geospatial big data is coming from every imaginable source, often in real time. Geospatial big data brings together not only large amounts of data but also various data types that previously never would have been considered together. These geospatial big data streams require speedily processing, economically storing, and timely feed-backing into decision-making processes. Traditional data models, databases, and tools are incapable of handling complex geospatial big data. Novel data mining algorithms that do not require all data to be loaded into the main memory for computing and effective computing tools and platforms that can take distributed large scale data storage into consideration are needed. In addition, existing tools cannot work well to sophisticated geospatial big data analysis at the scale many users would like. Current GIS tools such as ArcGIS software support relatively sophisticated analysis. However, they are not designed to scale to datasets that exceed even a single machine's memory. Even tools, which are designed to scale such as Geospatial Hadoop, don't support methods out of their box. Finally, the geospatial big data not only introduces query efficiency problems, but also other problems such as quality, privacy, perceptual and cognitive problems. It becomes more difficult to communicate and represent large, complex, and varied domain geospatial big data to the end users. We have to improve statistics and machine learning algorithms and tools to deal with the complex geospatial big data and train students in their intricacies. We also need to develop a data management ecosystem around these algorithms and tools so that users can manage, visualize, and understand the results of these algorithms and tools for the geospatial big data.

Cloud computing, which has powerful computing ability, may be a good technology to process geospatial big data. Because data scale of geospatial big data is far beyond the capacity that a single personal computer can handle, a large number of cluster computers with high-performance computing platforms and a parallel computing infrastructure may be deployed for dealing with geospatial big data.

In the following section we will introduce Cloud computing and the challenges that we are facing using this technology for Geospatial Semantic Web.

7.6   Cloud computing issue

As aforementioned the existing inference engines on Geospatial Semantic Web cannot process mass ontology. The popular inference engines such as Jena, Pellet, and Racer cannot handle more than 10 million triples on a single PC. Some inference engines such as BigOWLIM can handle tens of millions of triples on a single PC, but the performance is very low. Oracle 11 g spatial can reason on 100 millions triples, however, it needs dozens of hours to reason them (Qu 2012). The performance of Geospatial Semantic processing can be improved by using cloud computing technology. The main idea of cloud computing is to make full use of conventional parallel data processing via cloud platforms thus the system can process mass data.

Cloud computing refers to the replacement of physical computers in institutions with servers that reside "in the cloud". Based on NIST (Mell and Grance 2011) Cloud computing is a computing technology that leverages cloud's resources for "enabling ubiquitous, convenient, on-demand network access to a shared pool of configurable computing resources (e.g., networks, servers, storage, applications, and services) that can be rapidly provisioned and released with minimal management effort or service provider interaction".

Cloud computing can be accessible by institutions on an as-needed basis through three service models: 1) Infrastructure as a Service (IaaS). In this model, the Cloud computing providers offer computers, either physical or virtual, to be "leased" and managed by the users. This is the most basic cloud computing model and the providers only manage physical hardware in this model. 2) Platform as a Service (PaaS). In this model, the Cloud computing providers offer computing platforms typically consumed by web servers, databases, and development tools. 3) Software as a Service (SaaS). In this model, the Cloud computing providers manage the infrastructure, computing platforms, and applications. This is the most complicated model.

There are four types of Cloud computing deployment models (Nandgaonkar and Raut 2014): 1) Public cloud/external cloud, which allows cloud computing as openly or publically accessible. 2) Private cloud/internal cloud, which is managed or owned by an organization to provide the high level control cloud services and infrastructure for maintaining the security and privacy. 3) Hybrid cloud or virtual private cloud model, which compromises both private and public cloud models where cloud computing environment is hosted and managed by third party but some dedicated resources are privately used only by an organization. 4) Community model, which allows the Cloud computing environment to be shared or managed by a number of related organizations.

Cloud computing technology is affecting and transforming current computer infrastructure and services. Within just a couple of years, it becomes one of the most influential IT trends. Several characteristics such as scalability, cost saving, higher resource utilization, business agility, device and location independence distinguish Cloud computing from the traditional computing infrastructure. Cloud computing has powerful and scalable computing ability and it is particularly well suited for applications with a variable workload during different time. The main advantages of the Cloud computing are the followings: 1) it reduces the overhead costs associated with purchasing expensive equipment and the resources needed to maintain it. Users pay for only what is needed, when it is needed. 2) With Cloud computing users can increase computing capacity or add new computing capacity on the fly without investing in new infrastructure, training new personnel, or buying new software. 3) Using cloud computing users can access anything that they want from anywhere to any computer without worrying about anything like about their storage, cost, management and so on. 4) Cloud computing overcomes limitation of the traditional server-centric computing infrastructure, which does not fully utilize the computing power and storage capability of client systems. While traditional hosted environment allocates resources based on peak load requirements, the Cloud computing can dynamically allocate resources. 5) Cloud computing moves computation to remote data centers via Internet based communication. It provides a large variety of services and resources based on user-demand across a multiplicity of devices, networks, providers, service domains and business processes (Nandgaonkar and Raut 2014).

Although Cloud computing has the aforementioned advantages, there are still many challenges to adopt Cloud computing over the Geospatial Semantic Web and there is still a long way for Cloud computing to prove itself based on users' trust level.

The first challenge is interoperability faced by Cloud computing. It is great to have data where you need it, when you need it. However, the data is of no use if it cannot be communicated across several systems at once. Cloud computing interoperability problems arise when different Cloud providers try to cooperate exchanging data and applications. These incompatibilities may be technical such as different formats, incompatible programming code such as Java-based or PHP-based, or semantic. Existing Cloud computing solutions have not been built with interoperability in mind (Sheth and Ranabahu 2010). The competition among the Cloud computing providers such as Amazon and Google makes them promote their own incompatible software and reluctant to use standards. There is no portability of data or software created by the existing Cloud computing solutions. With the existing solutions, users are usually locked to one single Cloud infrastructure, platform or service and have to use the proprietary software designs, data models and data storage structures. The lock-in problem affects medium and small enterprises entering the Cloud market.

Interoperability can remedy the lock-in problem and benefit both Cloud users and providers. Interoperability in the area of Cloud computing means "the ability to write code that works with more than one Cloud provider simultaneously, regardless of the differences between the providers" (CCUCDG 2010). In an interoperable Cloud environment users can compare and choose Cloud offerings with different characteristics and switch among Cloud providers whenever needed without setting data and applications at risk. Tim Berners-Lee, who invented the World Wide Web and coined the term "Semantic Web", argues that semantically linking data may be "the missing part of the vocabulary needed to interconnect computing Clouds" and thus solving Cloud interoperability problem (Cerf 2009). Currently there are several research initiatives focusing on addressing semantic interoperability through standardized data models and APIs.

The second important challenge that Cloud computing faced is the scalability issue. Cloud computing should manage resources in such as a way that a program can continue running smoothly even as the number of users grows. It is not just that servers must respond to thousands of requests per second. Cloud computing must also coordinate information from multiple sources, not all of which are under the control of the same organization. While scalability is the best solution to increasing and maintaining application performance in Cloud computing environments, vertical scalability remains one of the main technological challenges. There are two types of scalability: horizontal scalability (Scale out) and vertical scalability (Scale up). Horizontal scalability is scaling through adding more machines or devices to the computing platform to handle the increased demand. Vertical scalability means scaling the size of a server by resizing the server or by replacing that server to bigger one. Vertical scalability can handle most sudden and temporary peaks in applications. However, because most common operating systems do not support on-the-fly (without rebooting) changes on the available CPU or memory, it is difficult to support the vertical scaling in Cloud environment. Currently most Cloud computing environments use horizontal scaling for automatic scaling purposes. In the future, more studies should be done to implement a Cloud computing infrastructure with capabilities of both horizontal scaling and vertical scaling. With capabilities of both horizontal scaling and vertical scaling it can make the most efficient use of computing resources in the Cloud computing infrastructure.

In addition, like the big data, Cloud computing also faces the security issue. Security issue may be one of the most important reasons that prevent users from accepting Cloud computing. Without doubt, putting data or running software on someone else's hard disk using someone else's CPU cause concerns to many users. There are various security issues such as availability, confidentiality, privacy, and accountability (Armbrust et al. 2010). Different methods such as use of multiple cloud providers, cryptography, improving virtual machines have been proposed to resolve these problems. For example, Cloud computing provided by a single company may have the problem of single point of failure even the company has multiple data centers at different geographic locations. This is because the company may use the common software infrastructure and accounting systems or the company may even go out of business. Many users will be reluctant to migrate to cloud computing without a business continuity strategy for such situations. Multiple cloud providers may guarantee high availability and solve the single point of failure problem. In general, Cloud users face security threats from both outside and inside the cloud. Many of the security problems faced by protecting clouds from outside threads are similar to those already facing large data centers. Security responsibility for Cloud computing environments should be divided among potential cloud users and providers. For example, cloud users should be responsible for application-level security while cloud providers should be responsible for physical security. As a new phenomenon to revolutionize the way we use the Internet, one must be careful to understand the security risks and challenges posed in utilizing Cloud computing.

Despite the potential gains achieved by the Cloud computing, the aforementioned issues and challenges hamper the growth of it. Although Cloud computing has the potential to become a frontrunner in promoting a secure, virtual and economically viable IT solution in the future, it still has a long way to go to fully implement its potentials.

Chapter Summary Since Geospatial Semantic Web is still at the initial stage of research, there are many challenges in implementing a workable system to query, retrieve, integrate, and visualize heterogeneous geospatial data at the semantic level. This chapter introduces several challenge issues faced by geospatial community to implement the Geopsatial Semantic Web for different applications. These challenge issues include issues in service composition, natural language processing, ontology, performance, big data, and cloud computing.

Geospatial web service composition is the process of reusing and logically recombining existing geospatial web services into composite services to fulfill a user query by providing new functionalities that existing geospatial web services cannot provide alone. Despite progress has been made for semi-automatic or automatic geospatial web service composition, there are still many of challenges for semi-automatic or automatic geospatial web service composition in a distributed and dynamic Internet environment. More studies are needed to compose geospatial web services by simultaneously selecting atomic services and inferring the composition patterns among the selected services so as to ensure services are composed in the best way with regard to QoS aggregation.

The GeoSPARQL semantic queries may present a challenge to the general users who have no training on how to use GeoSPARQL or the spatial data set that they wish to query. An ideal Geospatial Semantic Web should allow end users to profit from the expressive power of the Semantic Web standards such as RDF and Geo-SPARQL while at the same time hiding their complexity behind an intuitive and easy-to-use interface. A natural language processing (NLP) interface may help to achieve this goal. However, there are many challenges in the development of the NLP interface because of the ambiguity of the natural language.

Ontology quality and ontology matching are important for Geospatial Semantic Web. However, ontology quality and ontology matching remain a challenge for development of Geospatial Semantic Web. Nowadays web sites are no longer static web pages serving content and images any more. They become more responsive, adaptive, and dynamic. There is inheriting uncertainty in the ontology creation, maintenance, and matching processes. The use of newer data formats (many of which are schemaless) makes it harder to use existing ontology creating, matching and alignment techniques for the current web applications. Under these conditions existing approaches for ontology creation, maintenance, and matching need to be modified and new perspectives for solving this problem need to be developed for the future.

Representing structured geospatial data in GML, WFS, RDF, Geo-SPARQL can result in inefficient data access. Literature shows that obtaining spatial information becomes very slow when query the geospatial information from large geospatial databases over the Internet. Although different parallel approaches have been proposed in literature, there are still issues that need further studies.

With the development of Internet and GPS technologies, the geospatial information and data is unbelievably large in scale, scope, distribution, heterogeneity, and supporting technologies. Geospatial big data is coming from every imaginable source, often in real time. Traditional data models, databases, and tools are incapable of handling complex geospatial big data. It is difficult to capture, store, search, share, transfer, analyze, and visualize the geospatial big data over the Geospatial Semantic Web. New methods and tools need to be developed in the future to resolve the geospatial big data issue.

Cloud computing, which has powerful computing ability, may be a good technology to process geospatial big data. Cloud computing technology is affecting and transforming current computer infrastructure and services. Within just a couple of years, it becomes one of the most influential IT trends. However, there are still many challenges to adopt Cloud computing over the Geospatial Semantic Web and there is still a long way for Cloud computing to prove itself based on users' trust level.

References

Anyanwu K (2013) A vision for SPARQL multi-query optimization on Map-Reduce. In: Data Engineering Workshops (ICDEW), 2013 IEEE 29th International Conference, 8–12 April 2013, p 25–26

Armbrust M, Fox A, Griffith R et al (2010) A View of Cloud Computing. Commun ACM 53:50–58

Barrett MA, Humblet O, Hiatt RA et al (2013) Big data and disease prevention: from quantified self to quantified communities. Big Data 1:168–175. doi:10.1089/big.2013.0027

Boral H, Alexander W, Clay L et al (1990) Prototyping Budda: a highly parallel database system. IEEE Trans Knowl Data Eng 2:4–24 http://www.cs.albany.edu/~jhh/courses/readings/boral. tkde90.bubba.pdf.

Buttenfield BP (2002) Transmitting vector geospatial data across the internet. Geographic Inf Sci, LNCS 2478:51–64

Castagna P et al (2009) A parallel processing framework for RDF design and issues. http://www. hpl.hp.com/techreports/2009/HPL-2009-346.pdf. Accessed 15 May 2015

CCUCDG (Cloud Computing Use Case Discussion Group) (2010) Cloud computing use cases whitepaper. http://www.scribd.com/doc/18172802/Cloud-Computing-Use-Cases-Whitepaper. Accessed 15 May 2015

Cerf V (2009) Cloud computing and the internet. http://googleresearch.blogspot.com/2009/04/ cloud-computing-and-internet.html. Accessed 15 May 2015

Cruz SAB, Monteiro AMV, Santos R (2012) Automated geospatial Web services composition based on geodata quality requirements. Comp Geosci 47:60–74

DeWitt DJ et al (1986) GAMMA - A high performance dataflow database machine. In: Proceedings of the 12th international conference on very large data bases, p 228–237. http://dl.acm.org/citation.cfm?id=671463. Accessed 15 May 2015

Groppe J, Groppe S (2011) Parallelizing Join Computations of SPARQL Queries for Large Semantic Web Databases. In: Proceedings of the 2011 ACM symposium on applied computing p 1681–1686 http://dl.acm.org/citation.cfm?doid=1982185.1982536. Accessed 15 May 2015

Karjalainen M, Kemp G (2009) Uniform query processing in a federation of rdfs and relational resources. In: proceedings of the 2009 International database engineering & applications symposium, p 315–320. http://dl.acm.org/citation.cfm?id=1620469. Accessed 15 May 2015

Laney D (2001) 3D data management: Controlling data volume, velocity, and variety. META Group. http://blogs.gartner.com/doug-laney/files/2012/01/ad949-3D-Data-Management-Controlling-Data-Volume-Velocity-and-Variety.pdf. Accessed 15 May 2015

Lutz M (2007) Ontology-based descriptions for semantic discovery and composition of geoprocessing services. Geoinformatica 11:1–36

Mell P, Grance T (2011) The NIST definition of cloud computing. http://csrc.nist.gov/publications/nistpubs/800-145/SP800-145.pdf. Accessed 15 May 2015

Michael K, Miller D (2013) Big data: new opportunities and new challenges. Computer 46:22–24

Nandgaonkar SV, Raut AB (2014) A comprehensive study on cloud computing. Inter J Comp Sci Mob Comp 3:733–738

Peng ZR, Zhang C (2004) The Roles of geography markup language, scalable vector graphics, and web feature service specifications in the development of internet geographic information systems. J Geogr Syst 6:95–116

Qu Z (2012) Research on semantic processing for internet of things based on cloud computing. Inter J Adv Comp Tech 4:339–346

Ravindra P et al (2011) An intermediate algebra for optimizing RDF graph pattern matching on map-reduce. http://link.springer.com/chapter/10.1007%2F978-3-642-21064-8_4#page-1. Accessed 15 May 2015

Sheth A, Ranabahu A (2010) Semantic modeling for cloud computing, part I & II. IEEE internet computing magazine 14:81–83

W3C (2004) OWL Web Ontology Language for Services (OWL-S). http://www.w3.org/Submission/2004/07/. Accessed 15 May 2015

Wiegand N, García C (2007) A task-based ontology approach to automate geospatial data retrieval. Trans GIS 11:355–376

Yue P, Di L, Yang W et al (2007) Semantics-based automatic composition of geospatial web service chains. Comp Geosci 33:649–665

Zhang C, Li W (2005) The roles of web feature service and web map service in real time geospatial data sharing for time-critical applications. Carto Geogr Info Sci 32:269–283

Zhang C, Li W, Zhao T (2007) Geospatial data sharing based on geospatial semantic web technologies. J Spat Sci 52:11–25

Zhang C, Peng ZR, Zhao T et al (2008) Transforming transportation data models from unified modeling language to web Ontology language. Transp Res Rec: J Transp Res Board 2064:81–89

Zhang C, Zhao T, Li W (2010a) A framework for geospatial Semantic web based spatial decision support system for digital earth. Inter J Digital Earth 3:111–134

Zhang C, Zhao T, Li W (2010b) Automatic search of geospatial features for disaster and emergency management. Inter J Appl Earth Obs Geoinfo 12:409–418

Zhang C, Zhao T, Li W (2013) Towards improving query performance of web feature services (WFS) for disaster response. ISPRS Inter J Geoinfo 2:67–81

Zhao T, Zhang C, Anselin L et al (2014) A parallel approach for improving Geo-SPARQL query performance. Inter J Digital Earth. doi: 10.1080/17538947.2014.904012

Index

Printed by Printforce, the Netherlands